Landscapes That Speak

Jo Coghlan · Huw Nolan

Landscapes That Speak

Terrain, Threshold, and Transformation Across Cinema and Culture

palgrave
macmillan

Jo Coghlan
University of New England
Armidale, NSW, Australia

Huw Nolan
University of New England
Armidale, NSW, Australia

ISBN 978-3-032-21756-1 ISBN 978-3-032-21757-8 (eBook)
https://doi.org/10.1007/978-3-032-21757-8

This Palgrave Macmillan imprint is published by the registered company Springer Nature Switzerland AG
The registered company address is: Gewerbestrasse 11, 6330 Cham, Switzerland

Preface

This book responds to a profound transformation in how life on Earth is organised and experienced. The twenty-first century is marked by accelerating climate disruption, ecological collapse, extractive capitalism, militarised borders, and pervasive toxicity. These forces do not operate abstractly they are lived through landscapes. Landscapes are where planetary change becomes material, sensorial, and unevenly distributed across human and more-than-human worlds. They register rising seas and burning forests, but also displaced communities, altered animal migrations, poisoned soils, and atmospheres thicken with chemical residue. To understand contemporary crisis, therefore, requires attending to landscapes not as passive settings but as active sites where power, memory, violence, and care are entangled.

This book insists that landscapes matter because they shape how humans and nonhumans coexist, suffer, adapt, and endure. Landscapes carry the imprint of colonial dispossession, industrial extraction, and infrastructural violence, while also hosting practices of resistance, survival, and repair. They are ethical terrains. Decisions about where to mine, dam, fence, pollute, abandon, or protect land are simultaneously decisions about whose lives are valued, whose are rendered disposable, and which forms of life are permitted to persist. By foregrounding landscapes as threshold spaces—zones of instability, transition, and becoming—this book provides a way of understanding how ecological and social crises unfold together, and why ethical responsibility must be rethought in relational, multispecies terms.

This book was written out of concern with dominant ways of narrating crisis. Public and scholarly discourse often frames environmental breakdown as sudden catastrophe or future apocalypse, obscuring the slow, cumulative, and ordinary nature of harm. Such narratives also tend to privilege human experience while marginalising the more-than-human lives—animals, plants, waters, soils, microbes, atmospheres—through which landscapes are constituted. At the same time, cultural texts are frequently treated as secondary to actual environmental processes, rather than as powerful forces that shape how crisis is imagined, normalised, or resisted.

This book was written to bring these dimensions together. It places real-world landscapes—nuclear exclusion zones, extraction sites, borderlands, toxic corridors, submerged valleys—into dialogue with cinema, television, literature, and popular culture. In doing so, it shows how landscapes operate as narrative agents: they teach, warn, seduce, obscure, and remember. The book seeks to move beyond description toward an ethical engagement with damaged worlds. It asks what it means to live responsibly in environments that cannot be restored to purity; how grief might function as knowledge rather than paralysis; how refusal can interrupt extractive futures; and how maintenance and care might become central political practices in the Anthropocene. Writing this book has been an attempt to articulate an ethics adequate to worlds that are already wounded yet still inhabited.

This book is written for scholars and students across the environmental humanities, cultural and media studies, geography, sociology, anthropology, animal studies, Indigenous studies, and political ecology. It is also intended for artists, activists, policymakers, and readers beyond the academy who are grappling with climate anxiety, environmental injustice, and the experience of living in unstable worlds. The book speaks to those who sense that something fundamental has shifted in how humans relate to land, and who are seeking conceptual tools for thinking through responsibility, care, and collective survival.

Above all, this book is for readers who recognise that ethical life in the Anthropocene cannot be grounded in mastery, extraction, or denial. It is for those willing to think of and with landscapes as living, communicative, and more-than-human assemblages. By foregrounding the mutual vulnerability of humans and nonhumans, and by tracing how harm and possibility are unevenly distributed across terrain, this book invites readers to reconsider how futures are made. It argues that ethics emerges not from abstract principles but from situated engagement, from listening to land, attending to damage, and refusing the narratives that render certain places and lives expendable. In a time of cascading thresholds, this book offers landscape as a way of seeing, knowing, and acting otherwise.

Armidale, Australia

Jo Coghlan
Huw Nolan

Acknowledgements We would like to acknowledge and thank Rachael Ballard from Palgrave Macmillian for her championing of this book. Many thanks and much gratitude go to Dr Shaun Wilson at RMIT for his scholarly support for the project ensuring it got off the ground and quietly cheering on this book from the sidelines. We also thank our mentors, Professor Lorna Piatti-Farnell, Academic Dean at SAE Creative Media Institute, Auckland and Emeritus Professor Donna Lee Brien from Central Queensland University for their support of our scholarship and enthusiasm for our research. We also acknowledge and thank our colleagues, especially Dr Lisa J. Hackett, at the Popular Culture Research Network (PopCRN), Australia's leading popular and material culture research network. Many of the ideas in this book came from the conversations, conferences, and research produced by PopCRN.

In many ways, this book may not have come about if Rachel Carson didn't write *Silent Spring* in 1962 and Joni Mitchell didn't write *Big Yellow Taxi* in 1970.

This book is dedicated to Beatrice, Eloise, Ozzie, Scarlet, and Jasper and to those who follow who bring respect to the land they walk upon, listen to, and who imagine its past, present, and future differently.

Competing Interests The authors have no competing interests to declare that are relevant to the content of this manuscript.

About This Book

Landscapes That Speak argues that landscapes are not passive backdrops to human action but active, communicative forces that shape meaning, memory, power, and possibility. Bringing together environmental humanities, cultural and media studies, geography, sociology, animal studies, indigenous scholarship, and political ecology, the book develops a sustained argument for understanding landscapes as living terrains through which contemporary crises are produced, narrated, normalised, and contested.

The book is organised around the core concepts of *terrain*, *threshold*, and *transformation*. Terrain refers not simply to physical ground but to the material, symbolic, and affective conditions through which worlds are lived. Threshold names the dominant condition of the present: states of suspension, instability, and transition produced by climate change, extraction, militarisation, and toxicity. Transformation is understood not as progress or rupture alone, but as an ongoing process through which landscapes, bodies, and relations are worn down, reconfigured, and sometimes remade. Across the book, landscapes are approached as narrative agents, sites that speak, remember, instruct, and demand ethical response.

Chapter 1 establishes the book's conceptual and methodological foundations. It introduces landscape as an ontological condition rather than a visual category, foregrounding situated, embodied, and relational ways of knowing drawn from Indigenous concepts of country, phenomenological and affect theory, and critical landscape studies. It outlines the five key landscape forms that structure the book—*liminal, wonderland, mythic, toxic, and extractive*—and situates cinema and popular culture as crucial sites where these landscapes are rendered perceptible, affectively charged, and politically meaningful. Chapter 2 focuses on liminal landscapes, examining spaces of suspension and uncertainty in both real-world sites and screen cultures. Nuclear exclusion zones, climate-threatened coastlines, engineered waterworlds, borderlands, detention regimes, and war ruins are analysed alongside films such as *The Day After*, *Threads*, *Beasts of the Southern Wild*, and *Sicario*. These landscapes are read as zones where sovereignty falters, time becomes distorted, and life persists under conditions of prolonged crisis, making visible the ethics of endurance and exposure.

Chapter 3 turns to wonderland landscapes, exploring how the uncanny, strangeness, spectacle, and hyperreality shape perception and power. Moving across literature, cinema, interactive media, and built environments such as theme parks, malls, and spectacular cities, the chapter shows how wonderlands enchant, discipline, and naturalise inequality while also offering counter-wonderlands that resist dominant imaginaries. Attention is given to dystopian wonderlands and to non-Western and Indigenous forms that unsettle the politics of spectacle. Chapter 4 examines mythic landscapes as terrains of deep cultural meaning and collective imagination. Mountains, forests, deserts, islands, and caves are traced across folklore, cinema, television, and popular culture, revealing how myth structures ideas of belonging, sacrifice, revelation, and control. Case studies range from fairy tales and horror cinema to *Game of Thrones*, *The Truman Show*, and *Shutter Island*, demonstrating how mythic landscapes continue to organise modern power relations, environmental ethics, and narratives of sovereignty.

Chapter 5 addresses toxic landscapes and the ecologies of modern contamination. Treating toxicity as atmospheric, temporal, and sensory, the chapter explores how chemical harm becomes domestic, ordinary, and difficult to perceive. Through analyses of cinematic soundscapes, residue, taste, and time, the chapter shows how toxicity reshapes embodiment, memory, and responsibility, and how popular culture renders slow violence perceptible in ways that policy and data often cannot. Chapter 6 focuses on extraction landscapes and petro-cultures, examining mining sites, dams, oil infrastructures, and petro-suburban life. These landscapes are read as spatial expressions of capitalist modernity that normalise sacrifice, produce particular kinds of subjects, and reorganise everyday life around energy dependence. The chapter highlights how extraction landscapes teach societies what is considered expendable and how resistance, refusal, and alternative futures emerge from within these terrains.

The concluding chapter, *Reading the Land: Memory, Power, and Multispecies Futures*, draws the book together by proposing landscape as method, epistemology, and ethical practice. It argues for a situated practice of landscape reading grounded in attentiveness, responsibility, grief, and refusal, and oriented toward multispecies futures rather than human exceptionalism. The book closes with an invitation to read landscapes carefully, to listen to what they carry, remember, and demand, and to recognise that how we engage with land is inseparable from how we imagine ethics, solidarity, and futures in a damaged but still living world. The book offers a framework for thinking with landscapes across cinema, popular culture, and lived environments. It shows that landscapes do not merely reflect the world; they actively compose it, shaping how crises are understood, endured, and contested, and how more just and livable futures might still be made.

December 2025

Jo Coghlan
Huw Nolan

Contents

CHAPTER 1

Terrain, Threshold, Transformation: Screened and Lived Landscapes That Speak

Abstract Landscapes are not passive settings but active forces that organise memory, power, and feeling. When understood as products of time, matter, and social struggle, landscapes become sites that "speak," carrying the traces of labour, dispossession, trauma, ecological change, and uneven governance. They do not simply host social life; they register it. Landscapes are dynamic negotiations among geological, ecological, economic, political, and cultural trajectories that rarely align neatly. Read this way, landscapes are events rather than scenes: they accumulate atmospheres and sedimented histories, absorbing slow violence, attritional harm, and unresolved loss, while also generating attachment, awe, and longing. This analysis is structured through three interlinked lenses: terrain, threshold, and transformation. Terrain names landscape as a dense assemblage of human and nonhuman agencies, including weather systems, microbial life, infrastructures, and toxic residues. Threshold frames landscape as an ontological condition where categories fail and worlds meet, from coastlines and suburban fringes to borders, extraction zones, and post-disaster environments. Thresholds also emerge through quasi-events and incremental transitions that reorganise life without spectacle. Transformation attends to landscape as durational change across material, symbolic, and affective registers, shaped by mobility, capital, climate, and crisis ordinariness, but also by practices of resurgence, repair, and multispecies reconfiguration. Situated within screen cultures, this framework shows how film and television actively organise landscape through visual

J. Coghlan and H. Nolan, *Landscapes That Speak*,
https://doi.org/10.1007/978-3-032-21757-8_1

and sonic regimes that can naturalise colonial and capitalist spatial logics, while also opening fissures where alternative sovereignties, relations, and futures become perceptible.

Keywords Landscape · Ontology · More-than-human geographies · Thresholds · Liminality · Environmental memory · Screen studies · Post-colonialism

Can landscapes speak? Do they feel? Can they remember? Landscapes are the terrain laid out before us, available to the gaze, stable enough to be named and mapped. But the moment we linger in these question the apparent solidity of landscapes dissolves. Landscapes are not neutral surfaces waiting to be described. They are assemblages of time, matter, affect and power. Traditional cartographies reduce landscape to topography, but we insist that landscape cannot be separated from the temporal and relational forces that animate it. In rejecting the idea of landscape as a fixed backdrop or inert container for social life, instead seeing landscapes as embodying multiple trajectories that are geological, ecological, economic, political and cultural, which unfold simultaneously and intersect unevenly over time (Massey, 2005). Landscapes are not a completed entity but a provisional meeting-point of processes that are always still in motion. A river valley, for example, cannot be reduced to its visible form. It is shaped by the slow movement of sediment following upstream deforestation, by the infrastructural labour of irrigation and dam building, by the seasonal movements of animals and birds, and by histories of settlement, dispossession and displacement that remain embedded in place even when they are no longer publicly acknowledged. These trajectories do not align neatly, rather they coexist, collide and accumulate, producing a landscape marked by tension, inequality and unfinished histories.

This relational understanding of space has ethical and political force. If landscapes are constituted through ongoing relations rather than sealed in the past, then they remain open to contestation, change and re-imagining. The landscape is never simply *there*, it is continually being made and remade through power, labour and memory. To read landscape relationally recognises it as an active negotiation between human and nonhuman forces, rather than as a passive setting upon which civilisation and history are written and performed. Landscape then is not an object but an ongoing, if uneven, negotiation. A suburban fringe records the speculative momentum of real estate booms, the stalled beginnings of half-built estates, concrete malls and asphalt carparks, the quiet persistence of remnant bushland, and the demographic movements of families seeking affordability at the city's edge. Its landscape is an uneasy compromise between aspiration, abandonment, ecology, and economic precarity. An open-cut mine reveals geological strata millions of years old, the violent interruptions of capitalist extraction, the labour of miners whose bodies bear its toll, the displacement of First Nations communities, and the technical dreams of rehabilitation projects that promise futures built on uncertain ground. The pit is not just an extraction site is it a narrative, speaking of a contested story cut into the land.

Landscapes are events rather than scenes (Casey, 1996). They are made and remade through lived temporality of how it feels to dwell, move, remember and desire. The landscape is not just formed, it forms us, drawing human life into rhythms that exceed intention. Landscapes remember what happens to them and within them (Coghlan, 2025). Landscapes do not endure as physical settings. They accumulate and retain the traces of bodies, atmospheres, experiences, losses, ceremonies, and violences, producing memories sedimented into terrain. Through repeated use, abandonment, ritual, and harm, land becomes thick with past presences, holding layered inscriptions of labour, grief, celebration, and dispossession that continue to shape how places feel, remember, are sensed and inhabited in the present (Casey, 2004; Ingold, 1993). These sediments of memory are not symbolic but material and affective, embedded in soils, ruins, scars, and atmospheres, and often made legible through absence as much as presence (Tilley, 1994). These landscapes frequently register forms of *slow violence*, attritional harms that unfold gradually and are unevenly remembered or recognised, binding memory to terrain through processes of environmental damage, erasure, and deferred witnessing (Nixon, 2011). Landscape is a mnemonic medium, where past events persist not as stable archives but as contested, uneven, and often unresolved layers of meaning (Nora, 1989).

Landscapes are suffused with emotions of fear, awe, attachment, and longing. They are not simply perceived, they are experienced through bodies, atmospheres, and orientations where memory and emotion intertwine. We carry landscapes with us as we move through them, and they, in turn, carry the imprint of our traversing (Yi-Fu Tuan, 1977). This shift from topography to temporality makes clear that landscapes are not passive. Soil, weather, infrastructure, machines, microbes, and memory each have agency in shaping what landscapes become (Latour, 2005). Landscapes are *more-than-human geography* (Whatmore, 2002) rather landscapes are multispecies encounters instead of human-centred panoramas. Landscapes hence are co-produced through the movements and agencies of animals, plants, rivers, microbes, abiotics, weather, and technologies. Landscapes aren't scenery arranged for human interpretation but are entangled lifeworld's in which humans are not the only participant. This re-imagining invites us to understand landscape as a dynamic field of relations, animated by forces and beings that exceed human design or comprehension.

Threshold: The Ontological Condition of Landscape

If landscapes are dynamic, constantly shifting, accumulating, eroding and unsettling then the landscape is not a boundary but a threshold. They are always in the midst of becoming, tentatively assembling themselves out of forces that do not naturally cohere. The threshold is the place where landscape discloses its true ontology. Landscapes are where categories fail, where time folds, where worlds meet, and where meaning is never fully secured. Michel Serres (1995) considers thresholds as turbulence, zones where distinctions collapse into entanglement and where the familiar splinters into the strange. A threshold is neither here nor there, neither presence nor absence, rather it is a thickened interval in which possibility and dissolution coexist. Landscapes materialise thresholds in countless ways: in the trembling line where the tide meets its limit; at the desert's frayed edge where scrub dissolves into dust; along the suburban fringe where the city's promise thins into emptiness; and at the fire-line where living canopy gives way to ash.

Thresholds are equally felt in affective and atmospheric registers, those situations where the world seems to hesitate before revealing itself anew. The hush inside a disused factory awaiting demolition; the unsettled air of a community after a flood where residents speak softly as if the landscape itself might overhear; or the vibration of militarised boundaries where the political tension is so palpable it becomes a kind of weather. These are not merely suggestive moods, they are thresholds that expose the vulnerability of the structures we depend on, thresholds that reveal how thin the membrane is between order and its undoing. Elizabeth

Povinelli's concept of *quasi-events* (2011) helps illuminate these subtler transitions. Landscapes rarely shift through spectacle alone. More often they change through incremental processes of erosion, salinisation, pollution or drought, whose slow accumulation renders them more formidable. Thresholds are thus temporal wounds marking where the future intrudes upon the present without fanfare. When a reef turns bone-white, when a delta subsides millimetre by millimetre, when a farm succumbs to salt creeping invisibly through groundwater, we encounter thresholds of a quieter, more devastating kind. These transitions often go unnoticed, yet they reshape entire ecologies and lifeways.

Marisol de la Cadena (2015) expands the threshold into ontological terrain. In the Andes, the mountain is not simply a geological formation but an *earth-being* whose agency comes into conflict with mining companies, state bureaucracies and global capital. Here, the threshold is not between land and law but between worlds: between those who acknowledge the mountain as ancestor and those who understand it as resource. These worlds do not meet on equal footing, and neither can they be collapsed into a single reality. The threshold becomes a zone of ontological diplomacy, a space where radically different forms of existence and meaning touch. Indigenous knowledge systems approach thresholds not as fractures but as relational interfaces. Bawaka Country and Yolŋu scholars (2016) remind us that Country is tides, winds, animals, ancestors, spirits, and people who are entangled in ongoing relationship. In this way, the threshold is where reciprocity is activated not between separate entities but across a shared field of life. A shoreline, for instance, is not a dividing line between land and sea but a living participant in Yolŋu law and kinship. The threshold becomes a relational density, a place where responsibilities, stories, and obligations accumulate. Thresholds emerge not as moments of transition but as conditions of co-constitution. Landscapes as negotiations among forces that are social, ecological, geological and spiritual and do not naturally align in Western thinking.

Thresholds reveal landscape not as a unified whole but as a plural, layered assemblage, always at risk of coming undone and always in the process of being remade. They expose the fragility of order and the possibility of otherwise. They make space for the emergence of ecologies, new solidarities, and new political imaginaries. The abandoned factory becomes a site for ecological regeneration, the floodplain becomes a commons of shared vulnerability, the border zone becomes a crucible for solidarities forged in precarity, the burned forest becomes a threshold of renewal where species reclaim the scarred earth. Most importantly, thresholds remind us that landscapes are never final. They remain permanently in negotiation between the past that shaped them and the future that presses against them. Landscapes occupy a space between the worlds that claim ownership over them and the histories that continue to haunt them meaning they are caught between exploitation and care, destruction and renewal, despair and possibility.

Transformation: Material, Symbolic, Affective

Transformation is the landscape's most fundamental gesture. It is not confined to spectacular ruptures of earthquakes, floods, and fires but is constantly at work in quieter, less legible ways: in the slow migration of a dune, the encroachment of saltwater into once-fertile soils, the steady cracking of concrete as plants force their way through, the incremental greening of post-industrial ruins. These material shifts do more than alter physical form; they reorganise relations, unsettle expectations, and recalibrate what a place is understood to be. Transformation, in this sense, is not an occasional episode but the ongoing condition of landscape, its continuous negotiation between persistence and change. Nigel Thrift (2008) foregrounds this by insisting that transformation is *affective* and *performative*. Landscapes change through gestures, rhythms, and atmospheres. Change is not imposed from above; it is enacted through inhabitation, through the accumulated micro-movements of bodies and things. Tim Cresswell (2006) reminds us that movements of people, goods, animals, seeds, and ideas are central to how landscapes transform, again reminding us that they are static or benign. Migration and forced displacement, seasonal labour, tourism, and the circulation of commodities all leave marks that slowly and sometimes harmfully reorganise terrain.

A landscape becomes an archive of mobilities and crossing, paths become roads, informal tracks become recognised routes, fences and checkpoints re-channel animal and human flows. To trace these changes is to read the politics of mobility written onto landscapes. Gilles Deleuze and Félix Guattari's notion of *becoming* (1987) treats transformation not as a move from one stable state to another, but as an open-ended process of de-territorialisation and re-territorialisation. Landscapes participate in these becomings as they are folded into new assemblages of industrial zones, conservation areas, or extraction frontiers each reconfiguring what bodies can do and where they can go. Julie Cruikshank's work with Indigenous communities (2005) shows that these transformations unsettle cosmology as much as geography. As glaciers retreat under climate change, it is not only ice that is lost but stories, ethical relations, and reciprocal obligations bound to these glacial beings are also destabilised. Transformation is symbolic, moral, and epistemological: it alters how the world can be known and what responsibilities it calls forth. A melting glacier is both an environmental event and a crisis of meaning.

Transformation, in this sense, is not confined to moments of rupture or spectacular change but is bound to what *cruel optimism* considers as *crisis ordinariness*, the condition in which structural crisis becomes woven into the fabric of everyday life. Under crisis ordinariness, worlds do not suddenly collapse, rather they wear out slowly, as people and places adapt to conditions that are damaging yet normalised, ongoing, and presented as inevitable (Berlant, 2011, 95–100). Transformation unfolds as attrition rather than event, marked by incremental changes, compromised attachments, and the gradual erosion of livable futures. Landscapes shaped by extraction, toxicity, or abandonment exemplify this process. They persist, function, and are inhabited even as their capacity to sustain life is diminishing life. In this way, transformation names not progress or renewal, but the durational management of damage, where survival becomes the dominant mode of relations (Berlant, 2011, 101–103). Monocrop farming landscapes exemplify transformation through repetition and depletion. Soil degradation, biodiversity loss, chemical runoff, and aquifer exhaustion unfold incrementally, stabilised by seasonal rhythms and productivity metrics. These landscapes continue to function, producing food, employment, and profit, even as their ecological resilience erodes. Transformation then is the slow recalibration of what counts as health land, aligning with Berlant's account of attritional crisis managed through adaptation rather than repair (Berlant, 2011, 101–103). Along the petrochemical "Cancer Alley" corridor between Baton Rouge and New Orleans, everyday life unfolds amid chronic exposure to industrial emissions, contaminated air and water, and elevated cancer risks. There is no single moment of disaster that marks the landscape as broken. Instead, harm accumulates gradually through routine exposure, regulatory exemptions, and the normalised presence of refineries and chemical plants embedded in residential space. Communities continue to live, work, worship, and maintain social ties even as illness, premature death, and environmental degradation become statistically predictable features of place. This is precisely the conditions of cruel optimism and crisis ordinariness: a world that does not collapse but slowly wears out, sustained through adaptation to conditions that are structurally damaging yet treated as inevitable (Berlant, 2011).

Transformation in this register describes the recalibration of what counts as livable and what forms of damage are absorbed into the everyday. At the same time, transformation can be generative and resistant. Glen Coulthard (2014) argues that land-based practices such as returning to customary burning regimes, reviving food-gathering practices, or re-establishing community on ancestral territories transform subjectivity and social relations, enabling forms of self-determination that exceed colonial recognition. A reclaimed wetland, a re-established midden site, or a revitalised language linked to place does not merely alter the visible landscape; it creates different ways of being a people in relation to land. Karen Barad's (2007) notion of *intra-action* helps to name the profound entanglement of material and symbolic change. For Barad, entities do not pre-exist their relation, they emerge through relational encounters (2007). Applied to landscape, this means that transformation is not something that happens to a pre-given terrain, rather landscapes themselves come into being through the changing configurations of human and more-than-human forces. A river re-routed by engineering, for instance, is not a modified version of the old river but part of an entirely different assemblage of meanings, risks, and possibilities.

Anna Tsing's work on *patchy* Anthropocene landscapes (2015) further underlines how transformation is uneven, contingent, and multi-scalar. Clear-cut forests, derelict plantations and abandoned industrial sites become zones where unexpected multispecies collaborations arise, from fungi to scavenger species, creating new but precarious ecologies. Transformation is neither simple decline nor straightforward recovery. It is a reconfiguration of life in damaged but not lifeless places. Transformation also works through perception and the sensible. Jacques Rancière (2004) reminds us that the distribution of what can be seen, heard, and felt is politically organised. Land-use changes, heritage designations, and planning narratives all transform landscapes by altering how they appear and to whom. A valley marked as a sacrifice zone for extraction is transformed in policy long before the first machine arrives, conversely a landscape declared protected is transformed in law even as complex human and nonhuman histories persist beneath that designation.

These perspectives suggest that landscape transformation is an unfolding dialogue between matter, memory, and meaning. It is choreographed by ecological forces of storms, fires, tides, and species migrations as much as by cultural practices, legal decisions, capital flows, as well as memory and collective desires. Transformation destabilises the present by making visible its contingency: coastlines that seemed permanent slip away; climates once considered stable become erratic; infrastructures designed for one era strain under the demands of another. Yet transformation also gestures toward alternate futures. In post-industrial ruins where different ecologies take root landscapes host small rehearsals of other ways of living. Transformation is to recognise landscapes not as finished forms but as processes of *becoming*, processes that are always more-than-material, always entangled with symbolism, affect, and power. Landscapes don't just change, they change what can be felt, imagined, claimed, and resisted. They reshape the conditions of possibility for both human and more-than-human life. In this sense, transformation is not just one aspect of landscape, it is the means by which landscapes, and the worlds they sustain, are continually made and remade.

Situated Approaches: Country, Place, Embodied Knowledge

To speak of terrain, threshold, and transformation is to recognise that landscapes never appear the same to all who encounter them. They are understood through the bodies, histories, cosmologies, and relations that move through them. Landscape is not an object awaiting universal interpretation; it is always apprehended from a particular place and through a particular register. This is evident in Indigenous philosophies of land, which do not separate the knower from the known. In Aboriginal thought, Country is not a scenic backdrop but a sentient, knowledgeable entity whose rhythms, stories, and laws exceed human timescales. Deborah Bird Rose's (1996) depiction of Country as alive with memory and obligation articulates a truth echoed across Indigenous worlds where land is kin, relation, and teacher. Knowing is not an act of mastery but a practice of attention, reciprocity and care.

Other Indigenous scholars illuminate additional dimensions of this relational ontology. Kyle Whyte's (2018) work on climate injustice emphasises that environmental change is not an unprecedented emergency for many Indigenous communities rather it is a continuation of long histories of colonial disruption. In this sense, landscape becomes a register of cumulative harm and endurance. A river altered by industry or drought is part of a temporal storyline in which past dispossession shapes present vulnerability. Landscape is thus entwined with justice, kin-making, and the struggle to maintain continuity amidst imposed rupture. Eve Tuck (2009) foregrounds how landscape is implicated in land-based futures. For communities whose relationships with land have been systematically damaged or denied, landscape is a site for imagining and enacting decolonial alternatives. Even when land is inaccessible, fenced, or transformed beyond recognition, it continues to embody longing, resistance, and political desire. The landscape's significance persists in the forms of return, refusal, and resurgence that orient Indigenous futures, not as nostalgia, but as a forward-moving reclamation of relational life.

Extending this further is Tyson Yunkaporta (2019) view that knowledge is inseparable from pattern, place, and story. Landscapes encode complex systems that are ecological, social, and epistemic that can only be understood through forms of literacy grounded in humility and reciprocity. For Yunkaporta, place-thinking is a method of perceiving relational patterns rather than imposing external categories. This approach doesn't interpret landscapes, it participates in them. It affirms that all understanding emerges through the interweaving of mind, body, land, and collective memory. Such perspectives expose the limitations of colonial ways of seeing that cast landscape as empty, passive, or available for extraction. Much of what is represented as pristine wilderness is instead the afterlife of dispossession, where suppressed sovereignties linger in silence. These absences shape terrain as much as any visible presence. As Paul Carter (1987) has shown, colonial landscapes are built upon layers of forgetting and enforced quiet, making interpretation inseparable from historical reckoning.

Embodiment also conditions how landscapes are lived. As Arun Saldanha (2007) demonstrates, landscapes are racialised in their spatial logics; bodies endure different forms of scrutiny, threat, or belonging depending on how histories attach themselves to skin and movement. A path, a border, a field, or an urban street is not experienced uniformly but is encountered through atmospheres shaped by violence, privilege, and resistance; atmospheres that cling to terrain as much as to flesh. Christina Sharpe's account of *weathering* extends this insight, showing how structural forces saturate the very air that people move through, altering what it means to inhabit a landscape in the wake of historical oppression (2016). Astrida Neimanis likewise deepens the notion of weathering by foregrounding how bodies and environments are mutually exposed to climatic, chemical, and temporal forces. For Neimanis, weathering names the processes through which human and more-than-human bodies bear and absorb environmental conditions, carrying imprints of toxicity, care, and endurance (2017). These perspectives reveal landscape as something encountered not only through sight or narrative but through the body's own openings to the world, its vulnerabilities, its histories, its capacities to be affected and to affect in return.

Landscapes are also multiple rather than singular. A wetland might be experienced as habitat, sacred ground, a flood buffer, or a development site. These realities coexist not as competing interpretations but as distinct enactments of place that overlap. An anthropological understanding of multiplicity recognises that landscapes don't belong to a single world but function at the intersection of many worlds with their own logics, sovereignties, and obligations. This multiplicity extends to the more-than-human. Tides, winds, soils, insects, fire, spirits, and ancestors all contribute to how landscapes unfold. Understanding land as a multi-layered assemblage of agencies destabilises the assumption that humans stand apart from their surroundings. Instead, humans participate as one actor within an expansive and dynamic constellation. Knowledge emerges across these relations shared among humans and nonhumans alike.

Situated approaches therefore shift the very grounds on which landscape is understood. They reveal place as a dense and fragile mesh of relations, ecological, historical, political and spiritual meanings, which can't be separated from the particular lives entangled within them. Landscapes resist universal categorisation because they emerge through situated encounters of embodied experience, collective histories, environmental memory, and the continual negotiation of relations between human and more-than-human actors. To understand landscapes in this way is to understand that they are part of how life becomes meaningful, bearable, and possible. It foregrounds the ethical stakes of being in relation: landscapes shape us, respond to us, exceed us and depend on us in ways that call for attentiveness and responsibility rather than control. To think landscape through Country, place, and embodied knowledge is to enter a deeper conversation with the world, one in which we are required to listen differently, dwell differently, and imagine differently.

Landscapes as Living Terrains of Meaning, Memory, and Transformation

Landscapes have often been perceived as static visual backdrops to human activities or as aesthetic objects to be passively admired. We resist this by emphasising instead the symbolic and dynamic capacities of landscapes to shape and mediate cultural narratives, ecological relations, and collective imaginaries. Landscapes, as Denis Cosgrove (1984) compellingly argues, are not just seen, they are deeply encoded with meaning, inherently political and ideological, and actively involved in processes of identity formation and social memory. Tim Ingold (1993) asserts landscapes must be understood as continually evolving *taskscapes* actively shaped by the rhythms of human and nonhuman interactions and practices, making them dynamic assemblages rather than static scenes. Critically engaging with landscapes requires acknowledging their role as vibrant, living assemblages, complex interweavings of ecological relations, historical memories, ideologies, and multispecies interactions (Morton, 2010; Tsing, 2015). Multispecies ethnography explicitly positions landscapes as active mediators of coexistence, conflict, and precarious survival within the Anthropocene, where human and nonhuman lives are intertwined in profound and unpredictable ways (Tsing, 2015). Timothy Morton (2010) further enriches this by describing landscapes as ecological and cultural interconnections characterised by entanglement rather than purity or linearity, highlighting their inherently uneasy, paradoxical and complex nature.

Rob Nixon (2011) deepens this understanding even further by demonstrating how landscapes are shaped by *slow violence*, forms of environmental harm that accumulate gradually, often invisibly that disproportionately affect vulnerable populations across uneven geographies. Toxic waste sites, extraction, eroded coastlines, polluted groundwater, and industrial sacrifice zones exemplify how landscapes absorb long-term temporalities of damage. Nixon's framework situates landscapes as archives of delayed destruction, where harm unfolds over years or generations, making ecological devastation difficult to perceive yet impossible to escape. Lauren Berlant (2011) complements this with her concept of *slow death*, describing the attritional wearing-out of bodies under conditions of ordinary crisis and structural inequality. Landscapes of poverty, contamination, and infrastructural abandonment embody this slow wearing-down, linking environmental degradation with affective exhaustion, compromised health, and diminished futures. Berlant's framework highlights how landscapes do not merely signify crisis; they participate in processes that erode the capacity to live well. Slavoj Žižek's (2008) work on ideology and systemic violence provides another crucial dimension. Žižek argues that landscapes often naturalise or aestheticise the structural violence underpinning capitalist and colonial systems; extractive mining landscapes, manicured suburban developments, and tourist vistas all embed ideological narratives that disguise exploitation. For Žižek, the *ideological sublime* of nature imagery masks ecological collapse and political inequality, rendering landscapes complicit in the reproduction of dominant orders. Achille Mbembe's (2003) concepts of *necropolitics* further illuminate landscapes as terrains where sovereignty, violence, and mortality are spatially organised. Borderlands, occupied territories, refugee camps, prisons, and toxic industrial zones become landscapes of abandonment and exposure, where certain populations are rendered disposable or confined to death-worlds. These insights make clear that landscapes are not neutral environments but instruments of power that distribute life and death unevenly across populations.

Drawing on Lefebvre's (1991) insistence that space is never neutral but socially produced and structured by power, we treat landscapes as politically contested arenas shaped by historical struggle, ecological crisis, and cultural imagination. This interdisciplinary dialogue is enriched by Donna Haraway's (2016) concept of *sympoiesis*, or making-with, which positions landscapes as more-than-human worlds co-created through interactions between humans, nonhuman species and abiotic elements. Such an approach displaces the human from the centre, fostering instead an understanding of landscapes as collaborative, contingent, and continually emerging. Rosi Braidotti' (2013) posthumanist framing persuasively argues for the dismantling of entrenched binaries, such as nature/culture, human/nonhuman, and sacred/profane, in favour of more fluid, ethically grounded ecological and ontological entanglements, which further informs ideas of how the landscapes lives. In order to examine landscapes via these critical positions, we consider landscapes via interconnected thematic lenses of *liminal landscapes*, *landscapes as wonderland*, *mythical landscapes*, *toxic landscapes*, and *extraction landscapes* to illuminate modes by which landscapes function as potent thresholds, that rupture and transform.

Liminal Landscapes

Liminal landscapes are terrains where the world becomes strangely visible to itself in spaces of ambiguity, transition and haunting that interrupt the stability of the everyday. They sit at thresholds where the coherence of familiar categories falters, unsettling the binaries of urban/rural, natural/artificial, living/dead that ordinarily scaffold social and ecological life (Gordon, 2008; Turner, 1967). In these in-between zones, the seemingly solid ground of modernity softens and its promises become uncertain and its exclusions echo. Liminal landscapes reveal what dominant narratives attempt to suppress; histories that refuse erasure, ecologies struggling to reconfigure themselves, and communities living on the edge of economic, political, or climatic dissolution. The power of liminal landscapes lies in the way they render precarity perceptible. These landscapes include borderlands shaped by surveillance and ongoing contestation; abandoned urban peripheries where infrastructures decay; exclusion zones emptied of human life by toxicity and violence; and ghost estates that bear witness to speculative excess and economic collapse. They also encompass drowned towns preserved beneath reservoir waters, coastlines eroding under rising seas, and post-industrial ruins suspended between obsolescence and reinvention. Disaster-stricken regions trapped in prolonged recovery, refugee camps where displacement hardens into a cruel and enduring temporality, and demilitarised zones where the collapse of sovereignty is permanently inscribed into the land further reveal landscapes as sites of abandonment, endurance, and unresolved crisis. Each of these landscapes embodies and intensifies the condition of liminality.

What makes these landscapes profoundly consequential is that they expose the world's hidden architectures of power and vulnerability. Liminality is not random. It is produced through histories of colonial dispossession, through uneven development, through the abandonment of certain communities to environmental degradation, and through the prioritisation of profit over livability. In this sense, liminal landscapes function as diagnostic terrains, revealing where systems fail, where futures have been foreclosed, and where the burdens of crisis are spatially and socially distributed. They manifest the contradictions of late modernity. Yet liminality is not only a condition of loss but also a generative force. Because these landscapes confound normative structures, they open conceptual and political space for alternative forms of dwelling, imagining and organising. Liminal landscapes thus become laboratories of possibility, terrains where the potential for renewal flickers within the cracks of collapse. Their unsettled nature cultivates a capacity to think otherwise, to see futures not yet sanctioned by existing orders.

Wonderlands

The notion of landscapes as wonderland offers a critical lens for examining environments that are not only surreal or uncanny but fundamentally destabilising spaces where the real becomes unreliable and where the boundary between world and fantasy collapses (Baudrillard, 1983; Freud, 1919). These are landscapes engineered to seduce and unsettle in equal measure: dreamlike terrains in cinema, hyper-designed theme parks, digitally augmented cityscapes, and immersive architectural environments that overwhelm the senses. They are landscapes engineered materially, symbolically, or technologically to seduce and unsettle in equal measure. This dynamic is not confined to built or artificial environments but extends powerfully into natural landscapes that are framed, managed, and aestheticised through cultural mediation. In cinema, forests, deserts, oceans, and wildernesses are rendered as dreamlike terrains that invite awe while harbouring threat. They appear untouched yet are deeply shaped by extraction, conservation regimes, and historical violence. National parks, coastal zones, and scenic hinterlands are curated to produce immersion and wonder, even as they mask ecological precarity and human intervention. Across these contexts, natural landscapes operate as immersive environments that overwhelm the senses and the imagination. They blur distinctions between the organic and the engineered, the authentic and the staged, revealing how nature itself becomes a designed experience, one that captivates, disorients, and disciplines viewers and visitors alike.

They do more than provide escapist pleasure or spectacular distraction rather they perform cultural work by staging the deep contradictions of contemporary life; between fantasy and fatigue, enchantment and surveillance, consumer desire and the longing for authenticity (Caillois, 1961; Vidler, 1992). Wonderland landscapes reveal how cultural imagination is actively organised, managed, and contested. They make visible the affective and myth-making machinery embedded within landscape, showing how societies attempt to manage anxiety through spectacle, enchantment, and immersion, even as these strategies generate new forms of unease, disorientation, and instability. Wonderland landscapes script desire, orchestrate movement, and choreograph emotional experience, often rendering users complicit in the fantasies they consume. To enter such a landscape is to step into a zone where meaning is deliberately unstable. Wonderland landscapes draw attention to how cultural longing, nostalgia, fear and aspiration are materialised through spatial design. They amplify tensions already latent in everyday life: the oscillation between the hyperreal and the mundane, between the promise of limitless immersion and the exhaustion of being constantly addressed by spectacle. In this sense, wonderlands aren't escapes from reality but highly charged terrains that mediate reality revealing the fantasies that sustain contemporary culture and the instability that haunts them (Baudrillard, 1983). Far from trivial or ephemeral, wonderland landscapes teach us how societies negotiate the psychological pressures of late capitalism, how they aestheticise desire, and how they stage the very boundaries of the imaginable.

Mythical Landscapes

Mythic landscapes are terrains thick with symbolic resonance, cultural memory, and collective imagination. Across mythic traditions, they have long functioned as sites of trial, revelation, metamorphosis and moral testing. They are spaces where individuals and communities confront the unknown and return altered by the encounter (Jung, 1964). Their power lies in their elasticity, each landscape carries multiple symbolic valences that can be activated or subdued depending on historical moment, ecological crisis or political struggle, enabling them to speak across vastly different contexts. In the contemporary moment, mythical landscapes have acquired renewed potency as they are increasingly mobilised to negotiate ecological disruption, capitalist excess and the unresolved legacies of colonialism. The desert becomes a figure of exhaustion and extractive ruin, the forest a site of both refuge and threat, the mountain a terrain of sovereignty and contestation, and the island a condensed world where power relations crystallise. These landscapes operate as cultural diagnostics. They reveal the pressures, contradictions, and longings that shape the present, transforming ecological and political anxieties into spatial allegories that speak with mythic intensity. The symbolic energies of these terrains illuminate not only the fragility of the world as it is but the yearning for worlds otherwise. Myth provides the grammar through which contemporary culture articulates its fears of collapse, its fantasies of escape, and its tentative visions of renewal.

Yet mythical landscapes are not solely arenas for psychological projection or cultural anxiety. They are also powerful sites of reclamation and resistance. For Indigenous and postcolonial thinkers, the forest, mountain, desert, and island are not abstract symbols or empty canvases for narrative embellishment but living homelands, storied worlds, and kinship ecologies. These landscapes hold ancestral presence and epistemological depth. They refuse the erasures of colonial mapping and reassert the continuities of place-based knowledge. In such cosmologies, land is not a stage for human action but a co-agent in world-making. It is ancestor, teacher, and co-traveller: a living and storied being; an archive of law and memory; a network of relationships between human and nonhuman worlds. Myth does not represent landscape; landscape is myth. Stories, songlines, and genealogies do not sit atop terrain, they emerge from and return to it. Mythic landscapes in Indigenous thought cultivate obligation and reciprocity rather than heroic trial. This contrast foregrounds the political stakes of mythical terrains in settler-colonial contexts where Western mythic imaginaries of frontier narratives, empty wilderness, and heroic conquest continue to erase Indigenous sovereignty and modes of belonging. Mythic landscapes thus become terrains of contestation between competing ontologies: one that imagines land as empty, symbolic, or exploitable, and one that understands land as kin, ancestor, and the foundation of lawful existence. It is precisely this tension that grants mythical landscapes their contemporary potency. They reveal the unfinished work of decolonisation, the urgency of climate justice, and the cultural resurgence of Indigenous worlds that unsettle colonial fantasies of emptiness.

Toxic Landscapes

Toxic landscapes foreground terrains scarred by industrial ruin, extraction scarring, environmental degradation, and ecological contamination; they reveal the slow accretions of harm that accumulate across bodies, territories, and generations. These are landscapes shaped not by sudden catastrophe but by the incremental forces of systemic inequality, slow violence, and colonial legacies; the kinds of attritional damage that is omnipresent and structurally invisible. Toxicity does not simply contaminate environments; it reorganises worlds. It transforms soil, water, air, and bodies into archives of exploitation, rendering landscapes into repositories of chemical, historical, and affective residue. Whether in the form of radioactive exclusion zones, polluted estuaries, petrochemical corridors, abandoned mining pits, or the dispersed afterlives of electronic waste, toxic terrains operate as stark indexes of capitalist extraction and its enduring socioecological consequences. Such landscapes materialise the uneven distribution of vulnerability. They expose how class, race, and colonial power determine which populations are rendered disposable, whose lands are sacrificed, and whose futures are foreclosed. Toxicity reveals a politics of exposure: some communities live with the slow saturation of heavy metals, carcinogens, endocrine disruptors, or radioactive particles, while others benefit from the wealth extracted through their suffering. In this sense, toxic landscapes are not environmental conditions but expressions of structural violence, articulating the spatial logics through which domination is normalised and sustained. They mark the fault lines of racial capitalism, settler colonialism, and global manufacturing economies; systems that externalise their waste and risk onto those who lack the political power to refuse them.

Yet toxicity also has affective and ontological dimensions. These landscapes unsettle distinctions between animate and inanimate, living and dead, safe and dangerous. They generate atmospheres of uncertainty, dread, and resignation, but also of endurance and defiance. Toxic landscapes remind us that the boundaries between body and world are porous and permeable: contaminants traverse skin, flesh, soil, air, and water, implicating all beings within shared ecologies of vulnerability. Dark ecology further emphasises this unsettling intimacy, showing how toxic environments reveal the entanglement of all life within systems of harm, insisting that we confront the uncomfortable proximity between ourselves and the wastes and residues we produce. At the same time, toxic landscapes are not solely sites of despair. They are also terrains of resistance, refusal, and speculative re-imagination. Communities living amid toxicity often cultivate political solidarities and activist networks that challenge exploitative industries and state neglect. These landscapes become battlegrounds where environmental justice movements demand recognition, remediation, and reparative futures. Moreover, toxicity forces new modes of ethical and imaginative engagement: it asks us to reconsider concepts of purity, sustainability, and recovery; to confront the temporalities of contamination that stretch far beyond individual lifespans; and to imagine alternative ecologies of care, repair, and coexistence in damaged worlds. Toxic landscapes thus operate as potent cultural mirrors, reflecting the violences that underpin modernity, but they also open conceptual and political space. They illuminate the entangled vulnerabilities of human and nonhuman life, reveal the deep structures of inequality embedded in environmental degradation, and provoke the imaginative labour of thinking otherwise in the midst of ruin. They are not just wrecked environments; they are terrains where the unresolved contradictions of the Anthropocene become visible and where the struggle for more just ecological futures must necessarily unfold.

Extraction Landscapes

Extraction landscapes of mines, oilfields, refineries, and pipelines aren't just sites of resource removal but powerful terrains shaped by capitalist temporality, settler colonialism, and petro-modernity. Extractive landscapes reorganise land, bodies, atmospheres and futures, producing landscapes and bodies defined by absence, toxicity and dispossession. Energy humanities foreground this cultural dimension. *Petro-culture* as a totalising condition in which modern subjectivity, mobility, and aspiration are built around fossil fuels. Oil is not merely an energy source but an organising principle of everyday life, shaping infrastructures, labour, aesthetics, and national imaginaries. Extraction is also inseparable from colonialism. The understanding of settler colonialism as an ongoing structure rather than a completed historical event underpins analyses of how mining and drilling extend long-standing practices of land dispossession. Extractive landscapes are characterised as regions in which Indigenous territories, waters, and bodies are subjected to intensified exploitation under militarised and neoliberal regimes. Scholars of energy culture further demonstrate how sensory and affective attachments to petroleum bind communities to extractive industries even as these industries produce environmental harm and social vulnerability. Environmental justice research similarly shows that petrochemical corridors and mining belts reproduce racialised and classed inequalities through uneven exposure to toxins, infrastructural neglect, and environmental degradation.

Critical geographic approaches emphasise that extraction does not end at the mine mouth. Turning land into a resource requires extensive geopolitical, technological, and financial work, situating extractive sites within vast global networks of capital, logistics, and control. Contemporary extraction is deeply entangled with planetary supply chains, automation, surveillance, and militarisation, binding local environments to distant markets and political interests. Extraction also produces long-lasting toxic afterlives. Petrochemical residues move through soil, water, animals, and human bodies, blurring boundaries between environment and embodiment and revealing how harm circulates across more-than-human systems. The temporal dimensions of extraction are equally profound: phenomena such as climate change or radioactive contamination persist far beyond human lifespans, creating deep temporal disjunctions between economic gain and environmental consequence. Despite these enduring harms, post-extractive imaginaries continue to emerge. Relational and community-led approaches propose alternatives to extractive modernity, while land-based practices of resurgence articulate refusals of extractive worldviews altogether. Extractive landscapes therefore appear not only as sites of damage and dispossession but as arenas of ongoing struggle, where competing visions of life, value, and possible futures converge and collide.

Spectacle, Sovereignty, and the Screened Landscape

Although media industries of film, television, literature and music lean on landscape to ground their narrative universes, the spaces they render are never passive horizons. They are charged terrains through which meaning, memory, and ideology circulate. They are active cultural agents that organise meaning, mediating relations between story, spectator, and the broader imaginaries from which they emerge. To treat landscape as benign setting is to miss its function as a discursive formation: an assemblage of images, sounds, spatial logics and aesthetic conventions through which power, ideology, affect, and identity are produced and naturalised. Cinematic and literature landscapes are not benign objects to be represented but a medium of cultural power that shapes what can be seen, felt, and known (Mitchell, 2002). In the contemporary world, most people will never visit the deserts, forests, inner cities, borderlands or remote coastlines that organise their sense of the world. Instead, we learn these places through popular culture; through films, television, literature and music that circulate globally. We start from a simple but far-reaching proposition that much of the meaning-making of landscapes in the modern imagination comes from popular culture, and especially from screen media. Cinematic landscapes, we argue, do political and cultural work. They delineate who belongs, who governs, and which futures appear thinkable, while also holding the potential to unsettle those same boundaries.

The focus on cinematic and televisual representations of landscape reflects the distinctive capacity of film and television to render visible what is often obscured, normalised, or dispersed in everyday life. Screen media can visualise landscapes not simply as settings but as liminal spaces, wonderlands, mythical terrains, and toxic environments shaped by power, history, and extraction. Through aesthetic strategies, narrative structures, tropes, and visual rhetoric, cinema imagines what cannot easily be seen, measured, or narrated: environmental harm that unfolds gradually, unevenly, and away from centres of visibility. In doing so, film and television can amplify marginalised landscapes and make perceptible the experiences of the "environmentally dispossessed" (Nixon, 2011, 5), drawing attention to communities and ecologies rendered peripheral by dominant political and economic systems.

Cinematic representations are particularly effective in exposing the "corrosive transnational forces" of contemporary capitalism, including "petro-imperialism, the mega-dam industry, outsourced toxicity, and neocolonial tourism" (Nixon, 2011, 5). Where policy documents, scientific reports, and news media often struggle to convey the affective, temporal, and ethical dimensions of landscapes and their environmental harm, visual culture can register the lived consequences of these processes. Film and television can materialise the "temporal, geographical, rhetorical and technological displacements" endured by land and communities subjected to extraction and environmental degradation (Nixon, 2011, 7), making visible the uneven distribution of risk, damage, and responsibility across time and space. Rather than focusing on spectacular disaster cinema, our analysis centres on representations of landscape shaped by *slow violence*. Cinema and television are uniquely positioned to visualise the "violent geographies of fast capitalism" (Nixon, 2011, 7–8), tracing how extraction, infrastructure, and development remake landscapes long before crisis becomes visible or catastrophic. In this sense, screen representations do more than illustrate environmental harm; they offer critical frameworks for understanding how landscapes are produced, exploited, abandoned, and remembered under conditions of late capitalism.

Landscape, in this sense, is not only in the material world but it is our cultural worlds. Film, television, literature, and music, as popular culture, take landscapes and reworks them making them legible through its own visual and sonic regimes. Giuliana Bruno (2002) describes screen as a form of spatial thought that generates affective itineraries; spectators do not simply watch moving images, they travel through them. Cinematic space is not just representation but a sensory and emotional geography in its own right, shaping orientation, memory and embodied perception. Cinematic landscapes are also structured by racialised, gendered and colonial gazes. Laura Mulvey's (1975) analysis of visual pleasure shows how dominant cinematic forms produce controlling visions centred on a masculinised spectator; women are framed as objects, positioned to-be-looked-at rather than as agents of looking. A similar logic extends to landscape where terrain is frequently feminised, domesticated, or eroticised and often cast as passive surface awaiting cultivation, conquest or redemption. Edward Said's (1978) critique of Orientalist representation likewise illuminates how landscapes of the global South are exoticised and flattened, presented as timeless, backward or mysterious for Western consumption. In settler-colonial contexts, these visual habits intersect with the white possessive logics that Aileen Moreton-Robinson (2015) identifies when land is framed as empty, ownable and naturally aligned with white entitlement.

It is precisely this tension between erasure and endurance that renders cinematic landscapes critical objects of study and ground much of our discussion. They operate as hegemonic instruments, converting territory into legible images that consolidate specific orders of race, gender, class and nation. At the same time, they remain terrains of potential counter-reading where ecological entanglements, colonised histories and alternative geographies can be reinscribed. Landscape, in other words, is both the language of power and the ground on which that language can be contested. Popular culture is central here because it has the power to naturalise specific ways of seeing space. Narratives, cinematography, and design promise cartographic mastery, positioning the spectator as if above the world, confirming sovereignty through overview. Editing and camera movement choreograph how places should be traversed and valued, subordinating certain spaces to quick, disposable glimpses while granting others extended, lingering attention (Lukinbeal, 2005). Within this regime, preferred readings (Hall, 1980, 1997) are secured through codes that affirm authorised versions of history, citizenship, the body, sexuality, gender, and race. The *male gaze* that frames women as spectacle (Mulvey, 1975) also frames the land as spectacle, rendering it available for possession, improvement or rescue. Racialised and colonial gazes exoticise or pathologise non-Western or Indigenous spaces, translating difference into scenic otherness compatible with tourist consumption (Said, 1978; Urry, 2002).

Cinematic landscapes also routinely mobilise what John Urry (2002) calls the *tourist gaze*: a distant, elevated, cataloguing way of looking that converts place into experience and encounter into commodity. These representational habits align closely with a capitalist production of space. Cities, suburbs, deserts, mountains, and coastlines are packaged as exchangeable brands, their use value to local dwellers subordinated to exchange value for investors, audiences, and destination marketers (Harper & Rayner, 2010; Lefebvre, 1991). In Guy Debord's (1994) terms, their *spectacle* turns social relations into images. Screen landscapes become consumable icons whose circulation displaces the labour, extraction and governance that produced them (Debord, 1994). The result is an uneven visual order meaning some spaces are hyper-visible as sites of investment, aspiration and leisure, while others, often associated with racialised communities, feminised labour, or ecological sacrifice, are rendered marginal, dirty, or disposable.

Early Cinematic Landscapes

From early cinema onwards, these dynamics have been embedded in screen practice. Early Hollywood frequently treated landscape as space rather than place: studio backlots and painted sets provided generic environments that could be repurposed across films, with little reference to specific geographies or histories. A single street façade might pass as New York, Chicago, or an anonymous American town; landscape functioned as interchangeable container for plot. With the arrival of synchronised sound and the consolidation of the vertically integrated studio system, production increasingly moved indoors. Landscapes were painted onto backdrops or simulated in controlled environments, their flatness reinforcing the primacy of character and dialogue over spatial specificity (Lukinbeal, 2005). Landscape as place, where the narrative is anchored in the material and cultural character of a particular region remained an alternative but less dominant mode. When films do treat place seriously, the setting becomes an active element that shapes and constrains action, embedding the story within local histories, architectures and ecologies (Wollen, 1980). Early studio practices often treated landscape as decorative; an establishing shot or panoramic vista set the scene, but the film quickly shifted to interiors or medium shots where the narrative unfolded as though disconnected from the surrounding terrain. Landscape footage was frequently delegated to second-unit crews or drawn from stock libraries providing continuity without significantly shaping narrative causality, theme, or affect (Lukinbeal, 2005).

The logic persists in contemporary production. Urban environments are often treated as interchangeable façades, ready to masquerade as other cities. The practice of location masquerade, shooting Melbourne as Chicago, Vancouver as anywhere, or the Australian Gold Coast as generic tropical paradise underscores how landscape is still often approached as a flexible visual resource rather than a site with its own narrative agency. In such cases, landscape functions more as *mise-en-cadre*, a framing operation that determines what appears in the image, than as *mise-en-scène*, a force that actively shapes action and meaning. Framing, as Jean-Louis Comolli (1980) argues, is an ideological act: it constructs the viewer's relation to space, delimiting visibility and thereby conditioning what can be seen and imagined. The use of landscape as metaphor intensifies this ideological work. When deserts are coded as voids, forests as danger, suburbs as safety, or wilderness as redemption, landscapes don't just illustrate human feeling; they naturalise social and political narratives. As Duncan and Ley (1993) note attributing moral or emotional qualities to place makes cultural values appear as inherent properties of the land. Cinema repeatedly translates histories of conquest, extraction, patriarchy, and racial hierarchy into apparently self-evident spatial truths: the frontier as inevitable destiny, the city as corruption, the nation as landscape. In the United States, frontier and wilderness myths have long framed land as feminised and virginal, awaiting penetration and cultivation (Kolodny, 1975; Slotkin, 1992). Westerns, in particular, stage fantasies of white masculine authority, with landscapes serving as theatres for moral struggle in which Indigenous presence is either erased or rendered as obstacle (Tompkins, 1992). Such metaphors are not confined to classical cinema. Contemporary screen cultures continue to recycle and rework them. This is why popular culture is so central to landscape meaning-making: they don't just reflect existing spatial ideologies, they amplify and circulate them, embedding them in everyday viewing routines.

At the same time, landscapes in film and television are never purely pictorial. They are spatiotemporal processes, made and felt through movement, sound, and duration. Gilles Deleuze's (1986, 1989) distinction between the movement-image and the *time-image* offers a vocabulary for understanding this dynamism. In the *movement-image*, action orders space and time: landscapes are organised around practical sequences, shaping perception through sensory-motor continuity. In the *time-image*, action falters, images linger, fracture, or drift and duration itself becomes perceptible. Landscape becomes a site where time is felt, where an empty street, a still lake, or a receding glacier holds the viewer in a suspended encounter with change. Phenomenological and haptic accounts of spectatorship push this further. Vivian Sobchack (1992, 2004) argues that viewing is an embodied, reversible relation between spectator and screen; the body's motility and proprioception are implicated in how images of movement and space are registered. Bruno (2002) extends this into a theory of emotional mapping: films teach us to feel architecture and landscape through routes, thresholds, and passages. Laura Marks's (2002) concept of *haptic visuality* describes close framings, grain, and texture that invite tactile seeing, making terrain feel touchable, thick, and proximate. In such moments, spectators do not simply look at landscapes, they inhabit them virtually with their own sensory histories and political positions shaping the experience.

Sound is crucial to this inhabitation. Michel Chion's (2019) work on audio-vision shows that cinema's geographies are co-produced by sound and image; sonic cues anchor off-screen space, delineate acoustic horizons, and generate pathways through which spectators imaginatively traverse landscapes. Ambient noise, reverberation, and acousmatic voices extend space beyond the frame, creating atmospheres that shape how environments are sensed. Sound in popular culture teaches us what certain landscapes *feel* like: the constant hum of freeways in urban dramas, the metallic clang and subsequent eerie quiet of post-industrial zones, the rhythmic thrum of ceiling fans in humid interiors, the low drone of wind across deserts, the amplified crunch of snow in polar settings and the distant calls of birds or insects that index rural or wilderness spaces. These sonic cues do more than accompany the image, they construct atmospheric signatures that allow audiences to *inhabit* landscapes imaginatively, learning their textures, moods, and temporalities through sound as much as sight. These dynamics are especially visible in Anthropocene cinema, which foregrounds landscapes marked by extraction, contamination and climate change. Long-duration shots of deforested hills, burning peatlands, collapsing ice shelves or sinking delta cities invite viewers to dwell in slow, disturbing processes that exceed human pacing. Such images risk aestheticising disaster, but they can also generate a more intimate, haptic sense of environmental transformation (Fay, 2018; Marks,

2000). Anthropocene cinema thus reveals cinema's double role: it is both implicated in and reflective of the material systems reshaping the planet.

Across all these traditions, our consistent claim emerges. Cinematic landscapes are not neutral scenery. They are active systems of power. Through framing, movement, sound, and narrative convention, film and television organise space, time, and meaning. Popular culture is one of the primary mechanisms through which landscapes are made intelligible in modern societies, where belonging is assigned, exclusion normalised, extraction justified, and historical violence rendered invisible. We argue that cinematic landscapes function as ideological machinery that naturalises capitalist, colonial, racialised, and gendered orders, making them appear inevitable rather than constructed. Yet this machinery is never complete. Landscapes on screen are unstable and overdetermined, carrying residues of conflict, displacement, and ecological harm. Within their fractures and thresholds lie moments of refusal, spaces where dominant narratives break down and other relations to land, memory, and responsibility become possible. To analyse cinematic landscapes is to recognise how the screen's impressions of harmony and affect are scaffolded upon historical and geopolitical inequalities, and to search for the ruptures where this scaffolding gives way, allowing other stories, sovereignties, and futures to surface.

References

Barad, K. (2007). *Meeting the universe halfway: Quantum physics and the entanglement of matter and meaning*. Duke University Press.

Baudrillard, J. (1983). *Simulations*. Semiotext.

Berlant, L. (2011). *Cruel optimism*. Duke University Press.

Braidotti, R. (2013). *The posthuman*. Polity.

Bruno, G. (2002). *Atlas of emotion: Journeys in art, architecture, and film*. Verso.

Caillois, R. (1961). *Man, play and games*. University of Illinois Press.

Carter, P. (1987). *The road to Botany Bay: An essay in spatial history*. Faber and Faber.

Casey, E. S. (1996). How to get from space to place in a fairly short stretch of time. In S. Feld & K. H. Basso (Eds.), *Senses of place* (pp. 13–52). School of American Research Press.

Casey, E. S. (2004). *Remembering: A phenomenological study* (2nd ed.). Indiana University Press.

Chion, M. (2019). *Audio-vision: Sound on screen* (Revised ed.). Columbia University Press.

Coghlan, J. (2025, August 8). Rewatching *picnic at hanging rock* at 50: An unsettling portrayal of place, silence and disappearance. *The Conversation*. https://theconversation.com/rewatching-picnic-at-hanging-rock-at-50-an-unsettling-portrayal-of-place-silence-and-disappearance-260312

Comolli, J.-L. (1980). Technique and ideology: Camera, perspective, depth of field. In P. Rosen (Ed.), *Narrative, apparatus, ideology* (pp. 421–443). Columbia University Press.

Cosgrove, D. (1984). *Social formation and symbolic landscape*. Croom Helm.

Coulthard, G. S. (2014). *Red skin, white masks*. University of Minnesota Press.

Country, B., et al. (2016). Co-becoming Bawaka: Towards a relational understanding of place/space. *Progress in Human Geography, 40*(4), 455–475. https://doi.org/10.1177/0309132515589437

Cresswell, T. (2006). *On the move: Mobility in the modern Western world*. Routledge.

Cruikshank, J. (2005). *Do glaciers listen?* UBC Press.

De la Cadena, M. (2015). *Earth beings*. Duke University Press.

Debord, G. (1994). *The society of the spectacle*. Zone Books.

Deleuze, G. (1986). *Cinema 1: The movement-image*. University of Minnesota Press.

Deleuze, G. (1989). *Cinema 2: The time-image*. University of Minnesota Press.

Deleuze, G., & Guattari, F. (1987). *A thousand plateaus*. University of Minnesota Press.

Duncan, J. S., & Ley, D. (Eds.). (1993). *Place/culture/representation*. Routledge.

Fay, J. (2018). *Inhospitable world*. Oxford University Press.

Freud, S. (1919). The uncanny. *Imago, 5*(5–6), 297–324. https://doi.org/10.2307/1421445

Gordon, A. F. (2008). *Ghostly matters*. University of Minnesota Press.

Hall, S. (1980). *'Encoding/decoding', in Culture, Media, Language*. Hutchinson.

Hall, S. (1997). *Representation*. Sage.

Haraway, D. J. (2016). *Staying with the trouble*. Duke University Press.

Harper, G., & Rayner, J. (Eds.). (2010). *Cinema and landscape*. Intellect.

Ingold, T. (1993). The temporality of the landscape. *World Archaeology, 25*(2), 152–174. https://doi.org/10.1080/00438243.1993.9980235

Jung, C. G. (1964). *Man and his symbols*. Aldus Books.

Kolodny, A. (1975). *The lay of the land*. University of North Carolina Press.

Laura Mulvey 1975. Visual Pleasure and Narrative Cinema, *Screen, 16*(3), pp. 6-18

Latour, B. (2005). *Reassembling the social*. Oxford University Press.

Lefebvre, H. (1991). *The production of space*. Blackwell.

Lukinbeal, C. (2005). Cinematic landscapes. *Journal of Cultural Geography, 23*(1), 3–22. https://doi.org/10.1080/08873630509478229

Marks, L. U. (2000). *The skin of the film*. Duke University Press.

Marks, L. U. (2002). *Touch*. University of Minnesota Press.
Massey, D. (2005). *For space*. Sage.
Mbembe, A. (2003). Necropolitics. *Public Culture, 15*(1), 11–40. https://doi.org/10.1215/08992363-15-1-11
Mitchell, W. J. T. (2002). *Landscape and power* (2nd ed.). University of Chicago Press.
Moreton-Robinson, A. (2015). *The white possessive*. University of Minnesota Press.
Morton, T. (2010). *The ecological thought*. Harvard University Press.
Neimanis, A. (2017). *Bodies of water*. Bloomsbury.
Nixon, R. (2011). *Slow violence and the environmentalism of the poor*. Harvard University Press.
Nora, P. (1989). Between memory and history: Les Lieux de Mémoire. *Representations, 26*, 7–24. https://doi.org/10.2307/2928520
Povinelli, E. A. (2011). *Economies of abandonment*. Duke University Press.
Rancière, J. (2004). *The politics of aesthetics*. Continuum.
Rose, D. B. (1996). *Nourishing terrains*. Australian Heritage Commission.
Said, E. W. (1978). *Orientalism*. Pantheon.
Saldanha, A. (2007). *Psychedelic white*. University of Minnesota Press.
Serres, M. (1995). *The natural contract*. University of Michigan Press.
Sharpe, C. (2016). *In the wake*. Duke University Press.
Slotkin, R. (1992). *Gunfighter nation*. Atheneum.
Sobchack, V. (1992). *The address of the eye*. Princeton University Press.
Sobchack, V. (2004). *Carnal thoughts*. University of California Press.
Thrift, N. (2008). *Non-representational theory*. Routledge.
Tilley, C. (1994). *A phenomenology of landscape: Places, paths and monuments*. Berg.
Tompkins, J. (1992). *West of everything*. Oxford University Press.
Tsing, A. L. (2015). *The mushroom at the end of the world*. Princeton University Press.
Tuan, Y.-F. (1977). *Space and place*. University of Minnesota Press.
Tuck, E. (2009). Suspending damage. *Harvard Educational Review, 79*(3), pp. 409–427. https://doi.org/10.17763/haer.79.3.n0016675661t3n15.
Turner, V. (1967). *The forest of symbols*. Cornell University Press.
Urry, J. (2002). *The tourist gaze* (2nd ed.). Sage.
Vidler, A. (1992). *The architectural uncanny*. MIT Press.
Whatmore, S. (2002). *Hybrid geographies*. Sage.
Whyte, K. (2018). Indigenous science (fiction) for the Anthropocene. *Environment and Planning E, 1*(1–2), 224–242. https://doi.org/10.1177/2514848618777621
Wollen, P. (1980). *Readings and writings*. Verso.
Yunkaporta, T. (2019). *Sand talk*. Text Publishing.

Žižek, S. (2008). *Violence: Six sideways reflections*. Profile Books.

CHAPTER 2

Liminal Landscapes: Spaces of Suspension, Instability, and Becoming

Abstract Liminal landscapes are considered as spatial and temporal thresholds where ordinary structures loosen, categories blur, and worlds enter states of suspension, instability, and becoming. Building from van Gennep's rites of passage and Turner's reformulation of liminality and *communitas*, the chapter argues that liminal landscapes are not simply transitional backdrops but politically charged environments that actively shape how vulnerability, solidarity, and uncertainty are lived. Borderlands, disaster zones, ruins, detention sites, wilderness edges, engineered waterscapes, and toxic terrains operate as threshold worlds in which belonging and authority are renegotiated, often unevenly and violently. Liminality is extended beyond ritual to modern governance and environmental crisis. Postcolonial theory frames liminality as hybrid "third space" formation, while haunting studies locate it in the return of unresolved historical violence. Affective and performative approaches show how disorientation is felt through bodies, atmospheres, and repeated crossings of normativity. The chapter emphasises liminality as political architecture: states of exception and necropolitical borders produce enduring conditions of juridical suspension and differential exposure to harm. Environmental humanities further reposition liminality as planetary, with climate-driven tipping points, coastal margins, and slow-onset transitions producing landscapes where futures arrive as quasi-events rather than spectacles. Through case studies of nuclear exclusion zones, climate thresholds, dammed and submerged terrains, militarised borders, and conflict ruins, the chapter

J. Coghlan and H. Nolan, *Landscapes That Speak*,
https://doi.org/10.1007/978-3-032-21757-8_2

maps how liminality is made durable through infrastructure, abandonment, toxicity, and uneven histories of power. In doing so, it argues that cinema functions as a laboratory of liminality, intensifying these real conditions through frames that render slow violence, haunted time, and precarious sociality perceptible, while also exposing the fragile, contested possibility of otherwise worlds.

Keywords Liminality · *Communitas* · Borders · Hauntings · Necropolitics · Ruins · Environmental thresholds

Thinking about landscapes as thresholds is to consider places where change, uncertainty, and transformation occur. These are liminal spaces. The term liminality, from the Latin *limen* meaning threshold, describes more than a physical boundary. It conveys a state of suspension and instability, a moment when familiar structures or categories begin to blur. Liminality refers to situations in which the normal order of things loosens, identities shift, and new possibilities emerge. This sense of openness and ambiguity makes liminality a valuable way to interpret landscapes, real and imagined. It helps us understand why certain settings, deserts, ruins, motels, forests, or suburbs can feel both ordinary and extraordinary, fragile yet full of potential for transformation. Liminality has travelled far from its anthropological beginnings, becoming a critical framework for analysing instability, transition, and the reorganisation of meaning across social, political, and environmental domains. At its most concise, liminality names a threshold condition: a suspension of ordinary structures in which identities, relations, and orders become unsettled and open to transformation. Its power lies in conceptualising not only transition but the indeterminate temporalities, affects, and materialities through which change becomes possible.

Emerging from Arnold van Gennep's *Les Rites de Passage* (1909), liminality first described the middle phase of ritual transition, where participants are no longer what they were but not yet what they will become. Victor Turner's reformulation (Turner, 1967, 1974) gave the concept its contemporary reach: liminality becomes a mode of experience defined by ambiguity, vulnerability and potentiality. Turner's concept of *communitas* describes the intense, temporary bonds that emerge when individuals pass through liminal situations in which ordinary social hierarchies are suspended. In these moments, people experience a sense of shared vulnerability, equality, and collective becoming that stands in contrast to the structured roles and power relations of everyday life. *Communitas* is not a stable social form but a fleeting condition, arising in moments of transition, crisis, or ritual passage.

When applied to landscape, *communitas* helps explain how certain spaces like borderlands, wilderness zones, disaster sites, pilgrimage routes, protest camps, or abandoned terrains can momentarily disrupt established orders of belonging and authority. These landscapes function as liminal environments where normative rules loosen and new social relations become imaginable. The land itself is not merely a backdrop to this experience but an active participant, shaping how bodies gather, endure exposure, and negotiate risk, scarcity, and interdependence (Turner, 1969). Cinematic and cultural representations frequently mobilise such landscapes to stage encounters with *communitas*. On screen, harsh or marginal environments bring strangers together, flatten hierarchies, and expose shared precarity, even if sometimes only briefly. Yet these moments are fragile and contested. They may offer glimpses of solidarity and collective possibility, but they can also be recuperated, contained, or erased as order is reasserted. Reading landscapes through *communitas* therefore foregrounds their political ambivalence: they are spaces where alternative forms of sociality flicker into view, without any guarantee that they will endure.

Late twentieth-century theorists have again reconceptualised liminality as a structural condition of modern life rather than a discrete transitional stage. For Homi Bhabha (1994), the *third space* of postcolonial identity formation is irreducibly liminal: a site in which cultural meanings hybridise and unsettle colonial binaries. Avery Gordon's *sociology of haunting* (Gordon, 1997, 2008) likewise treats liminality as the register in which unresolved historical violences return, revealing the incompleteness of social orders. Sara Ahmed's *affective economies* (Ahmed, 2006) frame liminality as spatial and emotional disorientation, where affects shapes who feels at home or out of place. Judith Butler's theory of *performativity* (Butler, 1990) adds an ontological layer: gender is not stable identity but ongoing negotiation, a threshold repeatedly crossed through normative reiteration.

The political stakes of liminality become explicit in Giorgio Agamben's idea of the *state of exception* (Agamben, 1998), where liminality is transformed into a permanent governmental strategy as in the case of refugees and stateless people who are held in juridical suspension, neither fully inside nor outside the law or the state. Achille Mbembe's *necropolitics* (2003) extends this spatially: border zones, occupied territories, and militarised environments function as governed thresholds where populations inhabit states of slow death or conditional survival. Liminality becomes less a passage than a political architecture of exposure and precarity. Environmental humanities deepen this spatial reorientation. Nigel Clark (2011) emphasises that the earth is itself a dynamic threshold system of coastal margins, tectonic borders, and climatic tipping points that exceed human governance yet structure human life. Liminality is also corporeal. It is lived in the migrant's uncertain passage, the detainee's suspension, the flood survivor's altered terrain (Coghlan et al., 2005). Ahmed's *emotional geographies* emphasise that liminal states are sensorially charged by fear, anticipation, or estrangement. Christina Sharpe's *In the Wake* (Sharpe, 2016) offers one of the most penetrating accounts: Black life under racial capitalism exists in a perpetual liminality where the past remains active and the future foreclosed. Gloria Anzaldúa's *Borderlands/ La Frontera* (Anzaldúa, 1987) theorises *nepantla* as a psychic and cultural threshold, painful, destabilising, but generative of new epistemologies and resistant forms of being.

A landscape becomes liminal when it can no longer sustain a single coherent ontology, when it exists at the boundary between states, between times, between forms of life. These are landscapes that unsettle epistemology and unsettle subjectivity, disclosing what Tim Choy (2011) describes as *ecologies of comparison* where multiple ways of knowing place coexist uneasily. It is in these real-world terrains that cinema finds the symbolic and affective soil for its representations of liminality. Understanding these landscapes first allows us to grasp cinematic liminality not as metaphor, but as an aesthetic intensification of already-existing spatial and temporal conditions. Liminality's contemporary relevance rests in its analytic elasticity. It captures the precarious atmospheres of ecological crisis, the political production of populations suspended between rights and non-rights, the affective disorientation of global mobility, and the unstable identities shaped by postcolonial hybridities and performative negotiations. As both condition and process, liminality provides a vocabulary for describing worlds in flux; worlds marked by eroding coastlines, fortified borders, toxic infrastructures, and shifting forms of belonging. Its enduring force lies in revealing how power operates through thresholds, and how new cultural, political, and ecological possibilities emerge precisely where stability breaks down.

Nuclear Exclusion Zones: Life at the Edge of Uninhabitability

Chernobyl (Ukraine), Fukushima (Japan), and Maralinga (Australia) are quintessential examples of liminal landscapes caught between life and death, toxicity and regeneration, memory and erasure (McFarlane 2016, 167–182; Morris-Suzuki, 2014, 331–362; Nixon, 2011, 65–66). Nuclear landscapes embody a form of liminality that is both spatial and temporal: they are places where the future is contaminated, the past is frozen, and the present is a negotiation with invisible forces that exceed sensory comprehension. The Chernobyl Exclusion Zone is a stark manifestation of this: abandoned homes still hold children's toys; peeling wallpaper reveals earlier attempts at normalcy; and parks sit rusted. Time has stalled, but not entirely. Wolves, deer, and birds proliferate in the absence of humans, complicating distinctions between ruin and resurgence (Flynn, 2021, 90–110). This cohabitation of toxicity and life exemplifies what Deborah Bird Rose (2004) calls *entangled destinies*, where human and nonhuman futures are woven through damage, endurance, and ecological improvisation.

Fukushima intensifies this liminal ambiguity, as contamination permeates air, water, soil, and food systems, producing a disaster without stable borders. The boundaries of harm are diffuse and continually reconfigured by ocean currents, wind patterns, and shifting technological assessments of risk (Flynn, 2021, 96–95). As Tessa Morris-Suzuki (2014) argues, residents needed to learn to navigate this uncertainty through embodied practices; encountering the landscape not only as a site of scientific measurement but as an affective and sensory field in which danger must be interpreted without clear markers. What emerges is a terrain structured by *spectral risk*: danger that is materially present yet perceptually unstable, aligning with Sara Ahmed's (2004) account of *affective atmospheres* in which bodies register threat even when empirical confirmation is absent or contested. In this context, radiation functions as environmental haunting, a latent, temporally dispersed force whose effects accumulate across years and generations (Masco, 2006). Fukushima's post-disaster landscape is profoundly liminal, a space where the invisible circulates with the everyday and where residents continually negotiate the tension between habitation and harm.

Maralinga demonstrates with particular clarity that nuclear liminality cannot be separated from colonial power. The Australian desert functions simultaneously as a repository for radioactive waste and as an Aboriginal homeland ruptured by forced removal, secrecy, and state violence. As Brian McFarlane (2016) argues, Maralinga occupies a liminal position in the national imaginary, partially remembered yet structurally repressed, its toxic afterlives entwined with Australia's broader reluctance to reckon with settler-colonial harm. The land is thus doubly contaminated: materially by radiation and culturally by the ongoing dispossession encoded in the site's partial erasure from public memory. Christina Sharpe's concept of *wake work* (Sharpe, 2016) helps illuminate how Indigenous communities endure these layered devastations, living in a temporality where nuclear fallout intersects with the continuing wake of colonial violence. Such landscapes become sites where mourning, justice claims, and resistant forms of knowledge persist despite state narratives of remediation or closure. Taken together, Chernobyl, Fukushima, and Maralinga reveal liminality as durational, geopolitical, and ontological, fields in which human and nonhuman life must navigate the instability of worlds altered by contamination, abandonment, and uneven histories of power.

Climate-Driven Thresholds: Edges of Disappearance and Emergence

Climate change generates some of the most consequential liminal landscapes of the contemporary world. These are sites where the stability of place falters; coastlines slip away, glaciers retreat, deltas drown, and cities oscillate between dry ground and inundation. Liminality here is not a transient condition but an unfolding temporal and spatial suspension produced by anthropogenic forces. The Louisiana Bayous exemplify this slow, grinding transition between land and water. Rising seas, subsidence driven by oil extraction, dredged industrial canals, and petrochemical infrastructures unravel wetlands at rates that destabilise both ecological patterns and cultural identities (LeMenager, 2014; Maldonado et al., 2013). Houses once set firmly on land now perch above encroaching saltwater, cemeteries sink, roads disappear, and entire communities, especially Indigenous and Creole groups, negotiate forms of slow displacement that precede formal and often forced resettlement. These are, in Elizabeth DeLoughrey's (2019) terms, *tidalectic landscapes* spaces shaped by rhythm, repetition, erosion, and return, where the boundaries between land and sea become mutable, haunting, and historically saturated. The bayous are liminal precisely because they inhabit the edge of disappearance, revealing the precarity of habitation under late-modern capitalism and the uneven geographies of vulnerability.

Retreating glaciers mark another critical planetary threshold. Their melting does not simply record climatic shifts; it reveals temporal strata previously locked away. Melt zones expose ancient forests, ice-encased bodies, and geological archives, making glaciers sites where deep time erupts into contemporary political and ethical debates. Julie Cruikshank (2005) shows that for Indigenous communities in the Yukon, glaciers are animate, storied beings whose retreat disrupts cosmologies, relational ethics, and collective responsibilities. The loss of glacial presence is thus both ecological and epistemological: a destabilisation of knowledge systems and ontologies formed through long-standing human–ice relations. Melting glaciers become temporal liminalities, neither wholly ruins nor fully emergent ecologies, but threshold zones where planetary pasts collide with uncertain futures. Astrida Neimanis and Rachel Loewen Walker (2014) extend this analysis by conceptualising melting ice as part of *weather bodies*, entanglements of human and climatic processes that blur distinctions between organism, environment, and historical time.

Venice offers a further example of an urban threshold shaped by accelerating climate instability. Each *acqua alta* (high water event) transforms the city into a temporary liquid architecture where stone, water, and light rearrange spatial experience in unpredictable ways. Such events reveal, as Saskia Sassen (2014) posits that global cities increasingly confront environmental forces that exceed the regulatory and infrastructural capacities of the modern nation-state, generating new forms of exposure and fragmentation. Venice's liminality arises from its amphibious condition: it hovers between UNESCO world heritage site and looming catastrophe, between preservation and submergence. At the same time, as Kimberley Peters and Philip Steinberg (2015) suggest, water destabilises terrestrial assumptions of solidity and nation-bound fixity, rendering Venice a vivid case of how climate change produces hydrological liminality: a political and ontological condition in which governance, identity, and material infrastructure remain perpetually unsettled.

Engineered Waterworlds: Submersion, Memory, and State Futurities

Dammed landscapes such as the Three Gorges Reservoir in China and Lake Pedder in Tasmania constitute liminal terrains shaped by large-scale, state-driven hydrological transformations. These are landscapes where modernity submerges history, yet the past persists materially and affectively beneath the waterline. As scholars of political ecology note, mega-dams create *engineered thresholds*, producing spaces that oscillate between ecological vitality and ruin, displacement and development. The Three Gorges Reservoir is one of the most dramatic examples of what Henri Lefebvre (1991, 49–52) considers the production of *abstract space* in which the Chinese state reorganises lived environments into hydraulic infrastructures aligned with nationalist imaginaries of strength, modernity, and control. The magnitude of this project underscores its liminal character: the dam stands 185 metres high and 2.3 kilometres long, and its reservoir extends over 600 kilometres upstream, submerging approximately 632 square kilometres of land (Tilt, 2015). More than 1.2 million people were displaced, and 13 cities, 140 towns, and 1300 villages lie drowned beneath the surface (Magee, 2011). When drought lowers the water level, the tops of buildings, terraces, and temple foundations re-emerge, producing what Gordon (2008) terms *ghostly return*, the insistent visibility of past lifeworlds despite state attempts to overwrite them. This hydraulic remaking creates dual temporality: the imposed futurity of hydropower and the submerged, spectral persistence of older socio-ecologies.

Tasmania's Lake Pedder, though smaller in scale, represents an equally consequential transformation of landscape into liminal terrain. The inundation created a new impoundment covering nearly 240 square kilometres, with water depths reaching up to 30 metres, submerging the original glacial lake and its internationally iconic pink-quartzite beach (Richardson, 2025). What makes Pedder especially liminal is that the original lake remains physically intact beneath the reservoir as an ecological and cultural archive, produces a further *ecology of mourning*, wherein grief becomes a mode of environmental knowledge and a driver of political imagination. Pedder is thus a landscape suspended between loss and potential restoration; the call to restore Lake Pedder positions it as a political threshold, a space where drowned memory becomes a speculative horizon for ecological justice. These engineered waterworlds reveal that liminality is not merely an incidental outcome of environmental change but often the deliberate result of what Elizabeth Povinelli (2011) describes as late-capitalism's infrastructural governance. Large-scale dams enact a temporal politics: they promise futures of electrification, development, and national security. They create *haunted geographies* where submerged worlds linger beneath the surfaces of modernity, continually re-entering political discourse, cultural memory, and environmental activism. Engineered water landscapes therefore function as profound liminal sites, thresholds between presence and absence, between state futurity and ecological grief, between the visible and the submerged. Their spatial and temporal instability makes them central to understanding how modern infrastructures produce not only power and progress but also dispossession, memory, and the unresolved legacies lying just below the waterline.

Borderlands, Detention, and Displacement: Geographies of Suspended Life

Borders are not merely geopolitical boundaries, they are liminal landscapes where sovereignty is performed through the production of uncertainty, vulnerability, and suspended life. Contemporary border regimes extend far beyond territorial edges, transforming deserts, detention centres, jungles, seas, and riverine zones into spaces where life is held in abeyance. As Nicholas De Genova (2002) argues, the border produces asylum seeker and migrant illegality, creating populations who exist neither inside nor outside the nation but in a condition of perpetual precarity. The U.S.–Mexico borderlands exemplify this dynamic. Gloria Anzaldúa's (1987) concept of *nepantla* captures the psychic and cultural liminality of border living, while Jason De León (2015) demonstrates how the Sonoran Desert has been intentionally weaponised. Stretching across southern Arizona and northern Mexico, the Sonoran Desert, its liminality lies not only in its geography, situated between nation-states, climates, and ecologies, but in its transformation into a space where migration, sovereignty, and mortality converge, U.S. border policy has intentionally rerouted migrants into the desert's environmental extremity. Temperatures exceeding 45°C, minimal shade, vast distances between water sources, and disorienting topography produce what Sarah Horton (2016) names as a *border of dehydration*, a zone where exposure is structurally embedded into enforcement strategies. The Sonoran Desert is liminal because it holds migrants in a state between legality and illegality, recognition and erasure. Nicholas De Genova (2010) notes that the deaths occurring in this region are not incidental but symptomatic of a border regime that governs through attrition and abandonment. The desert embodies Mbembe's (2003) *necropolitics*, a form of governance that manages populations through differential exposure to death. Yet it is also a site of spectral presence with migrants' belongings, makeshift shrines, and unmarked graves forming what De León (2015) describes as a material record of journeys that states attempt to render invisible.

A similar spatial logic governs Europe's frontier zones, particularly along the Evros River between Greece and Turkey. The terrain forms a zone of governmentality where patrols and ambiguous jurisdiction expose migrants to drowning, disappearance, and legal limbo. The border river is a profoundly liminal environment: a threshold where sovereignty is unstable and survival is contingent. Further west, Calais constitutes a more urban form of liminal suspension. Michel Agier (2016) theorises its refugee camps as spaces of partial recognition and systemic exclusion where residents live in conditions shaped by long periods of waiting, regular police raids, and irregular humanitarian intervention. Didier Fassin's (2013) notion of *humanitarian confinement* is evident here: migrants are contained not through walls alone but through temporal stagnation. Beyond Europe, the central Sahara, particularly corridors through Niger and Libya, constitutes one of the world's most lethal liminal landscapes. Judith Scheele (2012) shows how overlapping sovereignties of state forces, smugglers, militias, and EU-funded security units produce migratory routes marked by coercion, abandonment, and disappearance. The Sahara thus becomes a necropolitical frontier where environmental exposure becomes a mechanism of governance.

In the Australian context, Nauru and Manus Island epitomise the extraterritorial extension of liminal governance. Located thousands of kilometres from mainland Australia, these offshore processing centres operate as deliberately remote carceral landscapes, designed to sever asylum seekers from legal protections, public visibility, and the possibility of stable futures. Scholars describe these sites as forms of *offshore sovereignty*, where the state displaces its border to foreign terrain in order to suspend rights (Mountz, 2011). The islands' isolation functions as a spatial technology: detainees are neither incorporated into the Australian polity nor entirely expelled, but held in an indefinite temporal void. Arguably, indefinite detention on Nauru and Manus constitutes a temporal punishment in which waiting becomes a form of state violence. The former Woomera Immigration Detention Centre in South Australia demonstrates the domestic counterpart to this offshore strategy. Remote, arid, and surrounded by expanses symbolically framed as empty, Woomera was a site where asylum seekers existed in what Agamben (1998) calls the state of exception: a condition where law is suspended, rights are withdrawn, and individuals are reduced to bare life. The desert amplifies this juridical uncertainty; the landscape itself becomes part of the carceral regime, producing what Mares (2011) terms *architectures of despair*.

War Ruins and Conflict Zones: Landscapes outside Linear Time

War ruins and unresolved conflict zones produce temporal liminalities in which the past persists so insistently that the present cannot fully consolidate. These are landscapes where destruction, memory, and suspended sovereignty coexist, generating *imperial debris*, material and psychic remnants that continue to structure social life long after violence appears to have ceased (Stoler, 2013). In such spaces, time itself becomes unsettled: neither wholly past nor fully present, but caught in a recursive loop of interruption. The Cyprus Green Line exemplifies this condition. Stretching through the heart of Nicosia, the buffer zone contains abandoned hotels, rusted military vehicles, shuttered shops, and collapsing homes precisely as they were left in 1974. This is a landscape frozen by geopolitical stalemate, a zone in which sovereignty remains unresolved and where futures remain indefinitely deferred. Povinelli's (2011) notion of the quasi-present is particularly apt here: the Green Line is a time that continues but does not progress, an environment in which political possibility is suspended. As Yael Navaro-Yashin (2012) argues, the material decay of this landscape creates a ruination effect that shapes everyday affect of dust, mould and broken infrastructure generating atmospheres of loss, uncertainty, and haunted sovereignty. The Green Line is thus a liminal terrain not only spatially but temporally, holding Cyprus in a state of unresolved becoming.

Sarajevo's Sniper Alley offers another powerful instance of conflict-induced liminality. The boulevard, once a central urban artery, still bears the imprint of the 1990s siege through bullet-pocked facades, reconstructed surfaces, and memorial plaques that punctuate the streetscape. The city's architecture forms what Eyal Weizman (2011) terms a *forensic landscape*, where buildings function as archives of violence, their wounds recording trajectories of bullets, explosions, and human movement. Sara Ahmed's (2004) notion of *affective economies* again illuminates how emotions circulate through such scarred spaces: fear, grief, and resilience become attached to surfaces, guiding how bodies move and orient themselves in relation to the city's traumatic past. To walk Sniper Alley is to encounter a spatialised memory of violence; an affective geography where trauma is both endured and resisted. These liminal war ruins exist between destruction and reconstruction, between mourning and renewal, between past violence and uncertain futures. They are neither fully alive nor fully dead, they are *rubble landscapes*, terrains that resist closure because their material remains continue to generate political and affective force (Gordillo, 2014).

The Korean Demilitarised Zone (DMZ) is one of the world's most striking examples of a liminal conflict landscape, a space where war has not ended but is indefinitely paused. Extending roughly 250 kilometres in length and 4 kilometres in width, the DMZ constitutes a vast territorial threshold separating North and South Korea. It embodies what Heonik Kwon (2010) describes as a politics of divided time, a temporal suspension in which the 1953 armistice halts hostilities without resolving them, leaving the peninsula in a condition of perpetual 'not yet' meaning the landscape is thus profoundly liminal, neither war zone nor peace zone but a juridical and affective in-between. What makes the DMZ exceptional is the coexistence of militarised tension and ecological flourishing. Decades of restricted human access have allowed rare species, wetlands, and forest ecologies to regenerate, creating an accidental wilderness. Yet this flourishing is haunted: the land is saturated with landmines, surveillance infrastructure, and militarised architectures. Such contested ecologies reveal how environmental health can coexist with political violence, producing landscapes in which life proliferates amidst technologies of death. The DMZ's paradoxical ecology underscores its liminality. It is at once a sanctuary and a battleground, a symbol of division and a reservoir of unexpected biodiversity. It functions as a temporal hinge, where the unresolved past of the Korean War continues to structure the present, and where potential futures of reconciliation, reunification, and further conflict remain suspended. Christina Schwenkel (2009) notes that in such post-conflict spaces, material ruins and militarised geographies produce *durational haunting*, a sense that the past is not simply remembered but actively lived through the landscape.

Cinema as a Laboratory of Liminal Landscapes

If real-world nuclear zones, climate frontiers, engineered waterworlds, borderscapes, and conflict ruins materialise liminality in concrete ways, cinema offers a parallel but distinct terrain in which these thresholds are imagined, rehearsed, and contested. Film does not simply depict landscapes; it actively produces *emotional and haptic geographies*, arranging space, movement, and texture in ways that orient spectators affectively within environments of risk, loss, or possibility (Bruno, 2002). In this sense, cinematic landscapes can be understood as laboratories of liminality: spaces where the instability of territories, ecologies, and sovereignties is intensified, stylised and made sensible. Hence, cinema is a key site where spatial imaginaries are forged and circulated. As Chris Lukinbeal (2005) notes, filmic landscapes mediate between the material world and the ideological work of narrative, translating deserts, ruins, coasts, dams, and border zones into visual regimes that naturalise some forms of belonging while rendering others precarious. John Harper and Justin Rayner (2010) similarly insist that landscape in cinema is a performative milieu through which histories of extraction, settlement, conflict, and displacement are encoded. The liminal geographies traced in the previous section, Chernobyl and Fukushima, Louisiana bayous and melting glaciers, Three Gorges and Lake Pedder, Sonoran and Saharan corridors, Calais, Nauru, the DMZ, reappear, often transfigured, in fictional and documentary cinema that grapples with toxicity, drowning worlds, walls, zones, and ruins.

At the same time, cinema has its own specific temporality and phenomenology of liminality. Vivian Sobchack's work on *cinematic embodiment* (Sobchack, 1992) suggests that films stage threshold experiences for viewers: we inhabit in-between positions as both here and elsewhere, safe and imperiled, witnessing and implicated. Eco-cinema scholars such as Scott Macdonald (2013) show how slow, durational attention to damaged or threatened environments produces a spectral sense of time akin to the *slow violence* and *durational haunting* that shape real landscapes. Thomas Elsaesser's writings on ruins and post-war European cinema (Elsaesser, 2005) likewise demonstrate how bombed cities and leftover infrastructures become visual figures for historical suspension and unresolved catastrophe. Cinematic liminal landscapes thus sit at the intersection of representation and world-making: they draw on existing geographies of crisis and governance while also shaping how those geographies can be seen, felt, and politically imagined. What follows is an examination of cinematic liminal landscapes: films that mobilise the aesthetics of liminal landscapes to think through, imagine and contest crisis, memory, and becoming. By turning to cinema, the analysis traces how these real-world thresholds are refracted into narrative form, how they generate particular genres and visual grammars, and how they invite audiences to inhabit, however briefly, the unstable edges of worlds in transition.

Cinematic Nuclear Liminality: Landscapes of Fallout, Haunting, and Suspended Time

Cinema has long been a privileged medium for exploring the psychic, ecological, and temporal dislocations of nuclear catastrophe. Film and television possess a unique capacity to render the invisible visible; to translate radiation, abandonment, contamination, secrecy, and fear into textures, colours, sounds, and rhythms. Through *mise-en-scène* nuclear landscapes are transformed into liminal spaces, territories suspended between life and death, presence and erasure, past and future. HBO's *Chernobyl* (2019) is one of the most accomplished televisual representations of nuclear disaster, distinguished by its dense atmospheric design and meticulous reconstruction of contaminated space. The series' visual palette of green-grey concrete, low winter light, particulate-filled air creates a world where danger permeates each frame. The *mise-en-scène* emphasises stillness: children's toys left in playgrounds, overturned prams in empty courtyards, dinner tables set for meals that will never be eaten, which function as visual metaphors for a life halted mid-gesture.

Chernobyl's liminality emerges through its treatment of time. Using long, slow panning shots ordinary life appears suspended, yet nature slowly reclaims the landscape: moss spreads, weeds push through pavement, birds perch on irradiated structures. This silent cohabitation of ruin and resurgence establishes the zone as neither dead nor alive, but in a state of perpetual becoming. Crucially, *Chernobyl* foregrounds radiation's perceptual elusiveness. Rather than using visual effects to depict radioactive presence, the series relies on auditory cues (Geiger counter clicks, ominous drones) and actor performance (the delayed onset of symptoms, the sudden collapse of firefighters) to evoke contamination as a creeping threat. The camera's lingering attention to dust particles drifting through shafts of light implicates the viewer in this atmosphere of invisible risk.

Through its formal strategies, *Chernobyl* transforms the exclusion zone into an *affective landscape*: a spatial field charged with dread, memory, and future harm.

Post-Fukushima cinema adopts a similar visual and narrative logic. Fukushima films dwell in a register of slow disruption, reflecting the temporality of a disaster that unfolds across years rather than seconds. In Atsushi Funahashi's *Nuclear Nation* (2012), the camera lingers on empty schoolyards, evacuated towns, and makeshift housing complexes. Long static shots emphasise absence as a primary aesthetic mode. The devastated landscape is captured not through dramatic ruin but through everyday urban emptiness. The *mise-en-scène* positions Fukushima as a space where normal routines have been stripped away but where the remnants of life remain on the surface, like residue. Ian Thomas Ash's two feature documentaries about children living in areas of Fukushima contaminated by the 2011 nuclear meltdown, *In the Grey Zone* (2012) and *A2-B-C* (2013) are two of the most significant documentary engagements with post-Fukushima life, distinguished by its focus on intimate domestic spaces, children's bodies, and the quiet routines of families negotiating invisible harm. The 2013 documentary's title references the medical test for thyroid abnormalities (A, B, and C classifications), immediately positioning the body, specifically, the child's body, as the primary landscape of nuclear exposure.

A2-B-C employs a direct cinema style, with handheld camerawork, minimal narration, and extended observational sequences. This aesthetic choice foregrounds a form of everyday liminality, capturing the ways mothers monitor radiation levels in schoolyards, children undergo repeated ultrasound examinations, and families negotiate whether to stay, leave, or return to contaminated areas. These scenes emphasise ambient risk: danger that permeates space yet remains imperceptible to the eye. The *mise-en-scène* reinforces this ambiguity. Classrooms look ordinary, yet the camera lingers on half-empty corridors, kitchens appear functional but Geiger counters sit beside rice cookers and children play outside, but their parents' anxious glances betray a profound uncertainty. These juxtapositions reveal a world in which the familiar has become estranged, not through visible devastation but through invisible contamination that transforms the mundane into a site of watchfulness and fear. One of the film's most striking strategies is its use of close-up examinations of children's throats, where doctors search for thyroid nodules possibly linked to radiation exposure. These images collapse the boundary between environment and body: contamination enters the frame through flesh, breath, and medical instruments. The body becomes a liminal interface, registering harm that the landscape conceals. This reflects the broader post-Fukushima condition, where environmental and corporeal exposure are inseparable. Crucially, *A2-B-C* renders visible the epistemological liminality of Fukushima: residents are caught between competing expert claims, shifting governmental statements, and the difficult work of making sense of risk without stable reference points. The film shows mothers comparing radiation readings, demanding transparency from officials, and documenting their children's symptoms. These scenes suggest that in post-nuclear landscapes, citizen science becomes a survival strategy, and uncertainty becomes a form of governance. In *A2-B-C*, liminality is not spectacular but quotidian. It is lived in schoolyards and kitchens, in medical clinics and playgrounds. Through this intimate cinematic lens, Ash reveals nuclear fallout as a social and sensory condition, one that reshapes everyday experience and redefines the meaning of safety, home, and future.

Released just 12 days before the actual Three Mile Island accident, were a Pennsylvania nuclear power plant malfunctioned leading to the partial meltdown of a reactor core releasing radioactive gas, *The China Syndrome* (1979) occupies a unique position in the nuclear cinematic canon. *The China Syndrome* achieved an uncanny synchronicity in which cinematic fiction and material nuclear crisis briefly occupied the same historical moment, intensifying its liminal aura. The Three Mile Island event in Pennsylvania became a cultural touchstone for nuclear risk, corporate secrecy, and technological fallibility, with its name often invoked in public debates, policy documents, and fictional representations, including the uncanny proximity of the 1979 release of *The China Syndrome*. The site remains a symbol of both the vulnerabilities and the regulatory challenges of civilian nuclear energy in the United States. The film constructs its liminal landscape almost entirely within the confines of the nuclear power plant. Its *mise-en-scène* is dominated by steel corridors, humming control rooms, blinking consoles, and thick lead doors, spaces designed for containment yet visually coded as vulnerable. Close-up shots of oscillating gauges, sweating technicians, and trembling hands create a sense of *temporal imminence*: disaster is always imminent but never fully realised. This perpetual "almost-event" positions the plant as a liminal space, hovering between stability and catastrophic failure. The thriller genre here reinforces liminality: suspense is built not on what happens, but on what *might* happen. The use of surveillance monitors and closed-circuit television within the diegesis foregrounds the film's central concern with visibility and secrecy. The characters, like the audience, can only access partial truths. The film's ending, open and unresolved leaves audiences suspended in uncertainty, mirroring the real-world ambiguity that surrounded nuclear safety debates in the late 1970s and the uncertainty of the Pennsylvania leak.

Mike Nichols's *Silkwood* (1983) shifts nuclear liminality from industrial landscapes to the intimate terrain of the human body. Karen Silkwood's body becomes a site of contamination, surveillance, and suspicion. The film's emphasises bodily vulnerability via harsh fluorescent lighting in the decontamination room, the cold sterility of laboratory spaces, the abrasive sound of alarms that signal radiation levels. Unlike films that depict external environmental ruin, *Silkwood* visualises nuclear danger as internalised. The body is the threshold where harm is registered, through coughing fits, dust residue glowing under ultraviolet light, and the repeated stripping and washing rituals of workers in decontamination scenes. These sequences highlight the liminal condition of workers exposed to contamination yet required to maintain productive normalcy. The film also employs domestic space to underscore liminality. Karen's cluttered kitchen, shared apartment, and everyday routines contrast sharply with the sterile corporate and laboratory settings, creating a narrative tension between ordinary life and the invisible threat permeating it. The result is a filmic landscape in which home itself becomes a threshold between safety and exposure.

The Day After (1983) and *Threads* (1984): Televisual Apocalypse and the Durational Ruin

American ABC's *The Day After* (1983) and the BBC's *Threads* (1984) represent two of the most harrowing televisual explorations of nuclear annihilation. Both employ a quasi-documentary style to convey the disintegration of social infrastructure, using familiar landscapes of suburbs, schools, and shopping centres as stages for catastrophic transformation. In *The Day After*, Kansas farmlands and Midwestern towns are framed through expansive aerial shots that render the landscape exposed, open, and defenceless; an ordinary rural world transformed into a target grid. The film's pre-blast sequences use warm natural light and wide depth of field to emphasise the recognisable textures of American pastoral life: grain silos, schoolyards, church steeples, and suburban cul-de-sacs. When the missiles strike, this visual stability collapses. Extreme high-angle shots and rapidly intercut montages of flattened houses, scorched fields, and overturned vehicles plunge viewers into a desolate, dust-filled environment where orientation becomes difficult and survival uncertain. The ash-laden light, stripped colour palette, and disorienting soundscape construct a liminal terrain in which the boundary between the living and the dead is unsettled. Because *The Day After* was broadcast as a network telefilm, watched by 100 million Americans, these images of ruin, exposure, and suspended life entered popular consciousness with unprecedented immediacy. The film's televisual reach transformed its post-apocalyptic *mise-en-scène* into a shared national imaginary, embedding nuclear liminality into collective memory.

The Day After emerged at a moment of heightened Cold War nuclear tension. Produced and broadcast in 1983, the film responded to escalating hostilities between the United States and the Soviet Union following the Soviet invasion of Afghanistan, the deployment of Pershing II and cruise missiles in Europe, and President Reagan's rhetorical framing of the USSR as an evil empire. The early 1980s also saw the revival of strategic doctrines such as limited nuclear war and first-strike capability, intensifying public fears about the plausibility of a nuclear exchange. Civil defence planning had largely fallen into neglect, leaving American audiences simultaneously anxious and uninformed about nuclear effects. The telefilm thus functioned not merely as speculative fiction but as a cultural intervention into a world characterised by arms-race acceleration and political brinkmanship. Its depiction of nuclear devastation, shown not in distant battlefields but in recognisable Midwestern towns, challenged official narratives that framed nuclear war as strategically abstract or geographically remote. In doing so, *The Day After* amplified the era's growing unease, contributing to national debates over deterrence, disarmament, and the ethics of nuclear escalation. The film's unprecedented reach made it a focal point in the broader public reckoning with nuclear vulnerability that defined the period.

Threads takes this further. Mick Jackson's *Threads* (1984) remains one of the most uncompromising depictions of nuclear catastrophe, distinguished by its refusal of narrative consolation or stable temporal framing. The film's power lies in its rendering of nuclear devastation as a prolonged liminal condition; a slow, chaotic unravelling of social, ecological, and bodily life rather than a single spectacular event. *Mise-en-scène* plays a central role. Sheffield's industrial skyline, terraced housing, and bustling public spaces anchor the narrative in a recognisable working-class cityscape. This familiarity heightens the rupture that follows the detonation, when Jackson reframes the same environments through wide shots of collapsed factories, scorched moorland, and claustrophobic interiors of makeshift shelters. Debris, broken masonry, and abandoned objects dominate the visual field, transforming the city into a rubble landscape that is neither fully dead nor remotely habitable. The film's cinematography and colour design intensify this sense of progressive unmaking. Bright, natural tones in the early sequences give way to ash-bleached palettes immediately after the blast, and later to muddy browns, greys, and sickly yellows. This gradual desaturation visualises the erosion of biological and social life, mirrored in lingering shots of barren fields, decomposing animals, and lines of refugees trudging across devastated terrain. Sound design reinforces the film's liminal atmosphere. Avoiding dramatic scoring, *Threads* deploys prolonged silences, faint sirens, static-heavy broadcasts, and the distant rumble of fires to evoke a world emptied of structure and meaning. This sonic minimalism signals the collapse of civil society: the continuing drone of official announcements echoes without authority into ruined spaces.

The film's temporal structure is equally destabilising. Spanning weeks, months, and years after the attack, *Threads* uses disjointed time jumps to convey the fragmentation of social rhythms and cultural memory. Seasons blur, calendars become irrelevant, and by the final scenes language itself has deteriorated and time becomes an unfamiliar, degraded dimension. Nuclear harm is also inscribed on the body, which the film presents as its own devastated landscape. With documentary-like restraint, *Threads* depicts radiation sickness as burns, hair loss, cataracts, and congenital deformities through close-ups that emphasise fragility rather than spectacle. The body becomes a narrative archive of exposure, embodying nuclear liminality at the most intimate scale. Jackson intensifies the film's documentary sensibility through voice-of-God narration, charts, survival instructions, and projected casualty estimates, creating a hybrid docu-drama form that exposes the inadequacy of Britain's civil defence strategies. These didactic interruptions serve as a kind of counter-cinema, revealing what state rhetoric obscures. *Threads* is itself a liminal cinematic form, part social realism, part horror, part public-information film. It unsettles genre expectations and viewer orientation, mirroring the social and temporal disintegration it depicts. The film's devastating final image of a stillborn, deformed infant held in a dim, makeshift environment collapses any remaining sense of futurity. *Threads* leaves viewers suspended in a world where hope, progress, and narrative closure have all been annihilated.

Threads arguable speak directly to Britain's Cold War history. The early 1980s saw escalating tensions between NATO and the Warsaw Pact, driven by the Soviet invasion of Afghanistan (1979), the stationing of U.S. cruise and Pershing II missiles on British soil (1983), and NATO's Able Archer military exercise, which brought the superpowers perilously close to misinterpreting war-game simulations as real offensive preparations. Domestically, the Thatcher government's unwavering commitment to nuclear deterrence, symbolised by its endorsement of Trident submarine-based missiles, provoked widespread public dissent. The British anti-nuclear movement surged during this period. The Greenham Common Women's Peace Camp (est. 1981) drew global attention as feminist activists blockaded missile silos and challenged the gendered logics of militarism and deterrence. The Campaign for Nuclear Disarmament attracted hundreds of thousands to its marches. Meanwhile, civil defence measures summarised in the widely ridiculed *Protect and Survive* pamphlets and animated public-information films were criticised for trivialising the un-survivability of nuclear war. Within this socio-political atmosphere, *Threads* functioned as a direct intervention. By depicting Sheffield rather than London as ground zero, *Threads* foregrounded how nuclear catastrophe would devastate ordinary urban and working-class communities. The film was not merely a dramatisation of nuclear war but a cultural critique of the assumptions underpinning Britain's nuclear posture.

Nuclear Liminality and the Cinematic Mapping of Crisis

Across these films and series, nuclear landscapes emerge not as singular catastrophic events but as cinematic articulations of liminality: suspended worlds where danger is invisible, time is unstable, and life unfolds in altered states. These nuclear texts share formal strategies of muted colour palettes, empty spaces, slow pacing, sustained atmospheric tension that reappear in cinematic representations of climate catastrophe, borderlands, and conflict ruins. This convergence matters. Nuclear cinema teaches viewers how to *read* landscapes of crisis: how to recognise danger without spectacle, how to inhabit uncertainty, how to interpret environments where harm is distributed unevenly and often invisibly. In this sense, nuclear liminal landscapes provide a visual grammar that climatological and migratory liminality increasingly adopt. Films about rising seas, melting glaciers, or weaponised deserts often deploy the same visual motifs, emptiness, suspended time, sensory ambiguity, that nuclear cinema pioneered. Nuclear cinema therefore serves as a conceptual bridge between different modes of planetary crisis. It reveals how contamination, abandonment, displacement, and suspended sovereignty can be visually rendered and narratively structured. It also underscores the shared aesthetics of *slow violence* across the nuclear, climatic, and border regimes we analyse in subsequent chapters. What nuclear cinema illuminates most powerfully is that landscapes in crisis do not simply change; they transform the very conditions through which cinema imagines, represents, and feels the world.

Cinematic Climate Liminality: Flooded Worlds, Melting Time, and Environmental Thresholds

If nuclear cinema imagines the world on the brink of contamination and temporal rupture, climate cinema maps landscapes in the process of drowning, drying, heating, fracturing, or disappearing. Films and television dealing with climate change frequently foreground liminal environments of coastal edges, glacial melt zones, floodplains, deserts, and atmospheric thresholds; places where the stability of the world slips and new ecological forms begin to emerge. In these texts, climate crisis is not merely backdrop but a spatial and temporal force that reorganises narrative, performance, and cinematic world-building. Climate films translate scientific abstraction into sensory experience. Like nuclear cinema, they mobilise the aesthetics of slow crisis, haunting, and threshold, but with an emphasis on ongoing transformation rather than catastrophic event.

The emergence of climate cinema as a distinct cultural formation is inseparable from the broader socio-political conditions that have shaped public consciousness of environmental crisis since the late twentieth century. Rising greenhouse gas emissions, accelerating ice melt, intensifying storms, expanding wildfire seasons, and the increasing frequency of once-in-a-century floods have collectively produced what Mora et al. (2013) describe as a *state of climatic departure*, a threshold at which historical baselines for weather, ecology, and habitability irreversibly shift (see also Jolly et al., 2015; Knutson et al., 2010; Steffen et al., 2015). Politically, the period since the 1992 Rio Earth Summit and the 1997 Kyoto Protocol has been marked by stalled international agreements, fossil-fuel lobbying, uneven national commitments, and the growing recognition that climate change disproportionately affects marginalised communities, Indigenous nations, and Global South populations. Social movements and Indigenous activism have reframed climate change as not only an environmental problem but a crisis of justice, sovereignty, and human rights. At the same time, media cultures have grappled with the slow, dispersed, and often invisible temporality of climate harm, conditions that challenge traditional narrative forms. The convergence of scientific alarm, policy contention, grassroots mobilisation, and sensory uncertainty has created a fertile ground for cinema and television to visualise climate crisis through liminal landscapes, representing a world in ongoing transition rather than discrete apocalypse. It is within this socio-political horizon that climate cinema has developed its characteristic aesthetics of inundation, drought, melting, drift, and suspended time.

Climate cinema repeatedly maps planetary instability onto liminal landscapes, rendering worlds in transition through shifting atmospheres, compromised ecologies, and suspended temporalities. In *The Day After Tomorrow* (2004), Manhattan's abrupt inundation and flash-freeze reconfigure the city into a threshold zone with its flooded stairwells and ice-entombed interiors visualising the moment modernity tips into uninhabitability. *First Reformed* (2017) internalises climate crisis through austere architecture and psychological unravelling, its symmetrical framing and frigid palette constructing a spiritual and temporal liminality in which ecological dread becomes a mode of inhabiting the present. Snowpiercer (2013) and its serial adaptation relocate climate collapse into a sealed, mobile microcosm. An endlessly moving train crossing a frozen Earth becomes a stratified liminal world, with each carriage operating as a distinct microclimate that stages the social and ecological consequences of enclosure. Occasional exterior shots of buried cities and frozen oceans reinforce a sense of global stasis, where movement persists but time itself appears immobilised. *Don't Look Up* (2021) shifts liminality to the discursive realm, converting newsrooms, digital platforms, and political rallies into chaotic in-between terrains where climate change denial displaces political and environmental urgency and the slow approach of the comet becomes a temporal threshold oscillating between spectacle and extinction. *Years and Years* (2019) dramatises climate crisis as episodic, cumulative, and unresolved, its climate crisis events filmed with documentary immediacy to foreground climate instability as an uneven disruption of everyday life.

Other films deepen this archive of climate liminality. *The Road* (2009) presents a world submerged in ash and colourlessness, its barren highways and skeletal forests forming a ruinous earthscape where life persists only as aftermath. *Interstellar* (2014) begins with a dust bowl America framed through dust-choked cornfields and shuttered schools, expanding climate liminality into cosmic registers through time dilation and broken chronology. *Take Shelter* (2011) positions looming storms and uncanny meteorological anomalies as psychic thresholds, blurring the line between hallucination and environmental warning through weather-saturated *mise-en-scène*. In *Aniara* (2018) climate displacement is extended into space, transforming a lost spacecraft into a drifting, sealed liminal world where time unravels and simulated Earthscapes collapse. *Chasing Ice* (2012) captures glacial calving through time-lapse cinematography that renders deep time visible, transforming melting ice into a temporal threshold where solid becomes liquid, and disappearance becomes spectacle. *The Burning Season* (1994/2008) depicts Amazonian deforestation as ecological borderwork, using aerial images of smoke plumes and charred clearings to show forests shifting into unstable, transitional zones. *The Impossible* (2012), centred on the 2004 tsunami, visualises extreme-weather liminality through turbulent water, overturned interiors, and disoriented bodies, aligning climate disaster with sudden spatial collapse. *Woman at War* (2018) reframes climate liminality through activism, situating sabotage within Iceland's windswept highlands, where vast volcanic plains and minimalist compositions produce a pastoral threshold between resilience and fragility. Across these works, climate change emerges as a condition of perpetual in-betweenness, a cinematic grammar of disappearing coastlines, melting ice, drifting vessels, destabilised skies, submerged cities, and fractured temporalities, through which film renders global crisis not as an event but as an unfolding reconfiguration of what landscapes, bodies, and futures can endure.

Beasts of the Southern Wild (2012) Climate Liminality, Tidal Landscapes, and the Aesthetics of Disappearing Worlds

Benh Zeitlin's *Beasts of the Southern Wild* is a significant cinematic articulations of climate liminality in contemporary global cinema. Set in the fictional Louisiana community called "The Bathtub," an unsubsidised, unprotected zone lying outside the state's levee system, the film constructs a world literally and metaphorically at the threshold of disappearance. Like *Threads* visualising nuclear ruin, *Beasts* uses formal strategies to articulate a landscape in which environmental instability, cultural precarity, and temporal uncertainty coalesce into a powerful vision of planetary crisis. The film's distinctive force emerges from its capacity to render climate change not as future catastrophe but as an ongoing condition of place, body, and social imagination.

The Bathtub is a quintessential liminal environment. Its geography of mudflats, wetlands, levees, submerged roads, and improvised dwellings places it between land and water, habitation and uninhabitability. Zeitlin populates this world with decaying boat hulls, rusting machinery, drifting debris, and houses on stilts, constructing a *mise-en-scène* that emphasises impermanence and porosity. Unlike apocalyptic films that depict abrupt rupture, *Beasts* represents slow inundation, where rising waters and eroding ground continually redraw the edges of the community. The Bathtub's interiors of trailers, shacks, and makeshift communal spaces are cluttered with objects that bear the marks of long use: tattered tarps, mismatched furniture, salvaged metal, hand-painted signs. These are not props but material testimonies to survival amid scarcity and constant environmental pressure. The *mise-en-scène* is saturated with evidence of life lived in transition, consistent with real-world bayou communities threatened by sea-level rise, subsidence, and petrochemical encroachment. This is a landscape that is already post-apocalyptic without ever having collapsed: a place where climate transformation is neither spectacular nor novel but structurally embedded in daily life. One of the films' most distinctive formal strategies is its commitment to low-to-the-ground cinematography, often from eye level. This framing choice produces a child's-eye view of the world, emphasising proximity to mud, water, vegetation, and animal life. The camera is handheld, restless, mobile and sensitive to tremors, wind, and the uneven movements of a landscape perpetually shifting. These techniques embed the viewer within the environment rather than placing them at a safe observational distance.

The film's colour palette oscillates from lush greens and golds in scenes of play, celebration, and community activity to stormy blues, greys, and browns as ecological threat intensifies. Water in particular becomes a chromatic force: murky brown during floods, shimmering under sunlight, slick with oil in post-storm contamination. This dynamic colour work visualises climate liminality as oscillation, between brightness and pallor, vibrancy and decay, growth and rot. The storm sequence uses rapid editing, close-up water droplets, flares of lightning, and disorienting motion to immerse viewers in the sensory chaos of extreme weather. Here, climate change is not an abstract condition but an embodied, haptic experience. Sound is central to *Beasts*' climate aesthetics. The film's aural landscape of buzzing insects, splashing water, groaning storm winds, and distant thunder creates an acoustic ecology that constantly asserts the presence of nonhuman forces. These environmental sounds are not background ambience but active participants in the film's narrative world. The swelling score, blending brass, percussion, and folk motifs, carries a distinctive emotional weight, oscillating between exhilaration and melancholy. It frames The Bathtub as both a place of community pride and impending loss. During storm scenes, however, the score recedes, replaced by the unsettling sonic presence of wind, rain, and cracking structures; a soundscape of environmental overwhelm in which human voices struggle to be heard. Like *Threads*, which uses silence to dramatise infrastructural collapse, *Beasts* uses shifting sound to emphasise ecological precarity recognising that the music of celebration cannot silence the grinding noise of storms.

Where nuclear films often visualise temporal rupture, *Beasts* articulates a tidal temporality, a rhythm of advance and retreat, growth and loss, ebbs and flows. Time in The Bathtub is cyclical rather than linear, shaped by storms, holidays, tides, and community rituals rather than clocks or calendars. This cyclical structure resonates with account of *tidalectic time*, in which coastal communities experience history through repetition and return. Floodwaters remain longer each year, storms arrive with less warning, and erosion progresses visibly. The narrative of The Bathtub is one of *slow violence*, damage unfolding gradually yet inexorably, undermining the prospect of sustained habitation. *Beasts* functions as the definitive cinematic articulation of climate liminality because it renders environmental crisis as an ongoing, embodied condition rather than a discrete event, treating the landscape not as backdrop but as an active agent whose tides, storms, and shifting boundaries shape the rhythms of life. The film's distinctive power lies in its fusion of mythic, ecological, and affective registers, allowing thawing aurochs, ancestral visions, and bayou ecologies to coexist within the same unstable world. It visualises place-based resilience amid environmental breakdown, refusing narratives of evacuation or technological salvation and instead grounding its characters in practices of survival, memory, and communal identity. Crucially, *Beasts* offers no closure: climate change remains a durational condition, a slow and persistent reconfiguration of space, time, and belonging. In this way, it occupies the threshold between disappearing and emergent worlds, making it a strong complement to nuclear liminality as explored through films such as *Threads* and *Chernobyl*.

Waterlands: Submerged Worlds, Dammed Futures, and Hydrological Liminality on Screen

If nuclear cinema imagines contaminated wastelands and climate cinema visualises atmospheric instability, water-centred cinema maps hydrological thresholds, landscapes suspended between water and land, habitation and abandonment, memory and disappearance. Films and television dealing with deltas, dams, rivers, flooded cities, and submerged ruins frequently foreground liminal landscapes produced by environmental change, infrastructural intervention, or colonial hydrology. These works translate the politics of water, from inundation to scarcity, engineering to dispossession, into spatial and temporal aesthetics in which water becomes both agent and witness of transformation. Cinematically, water operates not merely as setting but as a mobile, shape-shifting medium that unsettles boundaries, creates new geographies, and forces characters into states of negotiation and adaptation. Jia Zhangke's *Still Life* (2006) is one of the most compelling representations of dam-induced liminality. Set against the ongoing demolition and inundation caused by China's Three Gorges Dam, the film situates its characters within a half-vanished city of high-rises collapsing, workers dismantling entire districts, and water levels rising in the background. Jia's static long shots show buildings partially submerged, staircases ending in water, and rooftops exposed like islands. These images create a hydrological threshold landscape, where land is continually becoming water and the city appears both present and ghosted. The *mise-en-scène* foregrounds debris, scaffolding, and half-demolished interiors, making the urban fabric visibly transitional. The Yangtze River, widening and swallowing districts piece by piece, becomes a cinematic archive of what is being lost. Sound designs of distant demolition and lapping water, echoing emptiness and contributing to the atmosphere of suspended life. The film's slow pacing and documentary realism reinforce the sense of *slow disaster*, a transformation that unfolds too gradually to feel like an event but too forcefully to ignore.

Jennifer Baichwal and Edward Burtynsky's documentary *Watermark* (2013) traverses the globe to depict massive hydrological interventions of dams, irrigation systems, aquaculture, desalination plants, and the landscapes they transform. Aerial cinematography renders waterscapes abstract and monumental; river deltas resemble fractals, dam reservoirs appear as artificial inland seas, and irrigation grids generate geometric patterns etched into arid earth. These images place viewers within planetary liminalities, where water's presence and absence reorganise human life and ecological futures. The film's non-narrative structure allows viewers to inhabit these thresholds, neither fully natural nor fully engineered, while visual aesthetics emphasise scale, pattern, and the uneasy beauty of extraction. In the context of liminal landscapes, *Watermark* shows water as both vital and violently manipulated, flowing through political, industrial, and affective circuits that mark bodies and territories. The documentary *DamNation* (2014) focuses on the dismantling of dams across the United States, especially the Elwha and Glines Canyon dams in Washington State. Here, dams are represented not as achievements of progress but as infrastructural ruins, monuments to outdated technological futures and ongoing ecological harm. Sequences showing cracked spillways, silt-choked reservoirs, and ghost forests emerging from drained lakes depict a landscape in flux, neither river nor lake but something transitional. When dams are breached, time-lapse cinematography captures rivers cutting new channels through exposed sediment, creating dynamic liminal terrains where landscapes return to earlier states yet remain irrevocably altered. The film's political argument that dam removal enables ecological rebirth rests on this visualisation of water's capacity to reassert agency over engineered space.

The proliferation of large dams in the twentieth and twenty-first centuries has produced some of the world's most contested liminal landscapes; zones suspended between transformation and ruin, development and dispossession. Globally, more than 50,000 large dams have been constructed, many during periods of intense state modernisation, Cold War geopolitics, and postcolonial nation-building (Scudder, 2005; World Bank, 2000). These projects were frequently justified as engines of electrification, irrigation, and national progress, yet they generated widespread socio-ecological upheaval: the flooding of towns and farmlands, erasure of cultural sites, mass displacement of rural and Indigenous communities, and irreversible alterations to river ecologies. The political economy of dams, whether in China's Three Gorges Basin, India's Narmada Valley, the United States' Colorado River system, or Australia's Gordon/Pedder scheme, has thus been deeply entwined with what scholars identify as *hydraulic modernity*, the state-driven belief that landscapes can be reshaped at scale through technocratic engineering (Mitchell, 2002; Molle et al., 2009).

These schemes often reproduce colonial logics of territorial control, resource extraction, and Indigenous dispossession, exemplified by the displacement of Adivasi communities in India (Baviskar, 1995), the submergence of petroglyphs and ancient settlements along China's Yangtze, and the drowning of Tasmania's Lake Pedder despite national protest campaigns (Richardson, 2025). Increasingly, dammed landscapes are understood through the lens of *slow violence*, environmental transformations that unfold over decades, degrading sediment regimes, fisheries, biodiversity and water quality (Magee, 2011; Nixon, 2011). These slow-moving crises produce spatial and temporal liminalities, where submerged towns re-emerge during drought, where riverbeds oscillate between flood and desiccation, and where infrastructural futures promised by the state exist in uneasy tension with the haunting persistence of displaced pasts. It is within this socio-political terrain that films and documentaries representing mega-dams as ruined futures, extractive frontiers, or haunted hydrological monuments must be understood.

Borderlands: Cinematic Threshold Geographies, Suspended Lives, and the Aesthetics of Sovereign Edges

If nuclear cinema visualises contamination and climate cinema renders environmental instability, borderland cinema maps political liminality, territories where sovereignty blurs, rights dissolve, and bodies move through landscapes charged with risk, surveillance, and precarity. Borderlands are cinematic thresholds: deserts, rivers, fences, checkpoints, refugee camps, migrant shelters, and no-man's-lands where the spatial order of the nation-state becomes unstable. As visual environments, these borderscapes repeatedly stage in-between states, between legality and illegality, belonging and exclusion, safety and exposure. Cinema and television use these liminal terrains to dramatise how migration, detention, and state violence are inscribed into land, atmosphere, and the body, producing powerful representations of suspended life.

Contemporary border regimes emerge from decades of political hardening: the militarisation of the U.S.–Mexico frontier in the 1990s, the securitisation of EU outer borders after Schengen, the offshore detention architecture of Australia's Pacific Solution, and the weaponisation of deserts, seas, and jungles means these become deterrent landscapes. These borderscapes function not simply as lines but as zones of governance, extending hundreds and sometimes thousands of kilometres outward through surveillance technologies, detention infrastructures, and outsourced enforcement. They are landscapes which are zones of exception, where migrants and asylum seekers exist in prolonged states of legal and existential limbo. The Sonoran Desert, the Evros River, the Mediterranean Sea, Calais's makeshift camps, and Australia's offshore centres (Nauru, Manus) operate as spatial instruments of migration control, producing conditions in which survival itself becomes a test of legitimacy. These territories are not natural obstacles but deliberately engineered political liminal landscapes, where environmental extremity, isolation, and bureaucratic delay form part of a sovereign strategy of deterrence. In this political horizon, film and television turn borderlands into cinematic laboratories of liminality, visualising how the threshold spaces of migration expose the fragility of nationhood and the human cost of the border's uneven architectures.

Tony Richardson's *The Border* (1982) anticipates themes later developed in scholarly work on *necropolitics*. The desert is rendered through heat-haze cinematography, skeletal vegetation, and washed-out colour palettes, emphasising its hostile, threshold qualities. Migrants traverse these spaces at night, guided only by firelight, whispers, and the faint glow of distant cities, scenes that visually articulate the border as a liminal world where hope and death coexist. Landscapes of sand, broken fences, and abandoned personal objects gesture to a spatial archive of lives erased by policy. The desert becomes a deathscape; a liminal terrain that reveals the violence of border enforcement without showing it directly. Cary Fukunaga's *Sin Nombre* (2009) situates its narrative along the migratory railway routes through Mexica transforming trains and rail yards into mobile borderlands. The moving train becomes a liminal corridor suspended between nation-states, law, and danger. The *mise-en-scène* emphasises precarious bodies atop railcars, clinging to metal edges, or moving through forests and freight yards that are neither settlement nor wilderness. Long tracking shots follow migrants running alongside the train, visually mapping the movement between belonging and exclusion. The railway is at once refuge and threat, mobility and entrapment becoming a cinematic articulation of liminal migration. Alejandro González Iñárritu's *Babel* (2006) crafts a global borderland through its fractured narrative structure. The Moroccan desert, the U.S.–Mexico frontier, and urban Japan become interconnected thresholds where miscommunication and transnational vulnerability unfold. The Moroccan sequences use wide shots, intense sunlight, and sandy palettes to frame the desert landscape as one of misrecognition and sovereign overreach. In the Mexico–California crossing scene, the camera captures the border fence as both barrier and membrane, porous enough for passage, rigid enough for containment. *Babel*'s sound design, especially in scenes of disorientation and fear, intensifies the border's atmospheric liminality.

Ladj Ly's *Les Misérables* (2019) reframes the Parisian *banlieues* as urban landscapes where racialised policing, surveillance, and marginalisation produce everyday liminality. Though far from an international border, Montfermeil operates as a domestic frontier: a space where state authority is both hyper-present and dysfunctional. Handheld cameras and aerial drone shots trace the movement of youths through courtyards, stairwells, and abandoned lots, spaces that visually evoke containment and exposure. The film aligns with border cinema by showing how sovereignty produces liminality within cities as much as between nations. The Australian series *Stateless* (2020), based loosely on real events, situates border control within remote desert detention centres. The landscape of flat horizons, oppressive sunlight, and dusty winds emphasises isolation and sovereign detachment. The *mise-en-scène* of razor wire, temporary structures, fluorescent-lit interview rooms, and desert peripheries constructs a geography of *administrative liminality*, where detainees exist in prolonged temporal suspension.

Sicario (2015): Borderland Liminality, Sovereign Edges, and the Aesthetics of Suspended Violence

Denis Villeneuve's *Sicario* (2015) well articulates borderlands as a liminal landscape, rendering the U.S.–Mexico frontier as a juridically unstable, atmospherically charged zone where sovereignty fragments and violence operates through ambiguity. Like *Threads* visualises nuclear collapse and *Beasts of the Southern Wild* articulates climate transformation, *Sicario* maps a terrain in which the familiar logics of statehood dissolve, replaced by a set of spaces and temporalities defined by exposure, secrecy, and suspension. In Villeneuve's hands, the border is not a line but a cinematic threshold, a world in which everyday infrastructures including highways, checkpoints, tunnels, and desert expanses become architectures of uncertainty and dread. *Sicario* constructs the desert as an inherently liminal environment, neither fully empty nor fully knowable. The Sonoran landscape rendered through beige palettes, heat distortions, and distant mountain silhouettes functions as an index of vulnerability. Villeneuve repeatedly frames dust clouds, mirages, and hazy horizons, treating the desert as an atmospheric partner to the film's moral opacity. The frontier's built environments of chain-link fences, anonymous compounds, industrial warehouses, and checkpoints appear rudimentary and provisional, visually emphasising the unsettled nature of U.S. sovereign reach. Roger Deakins's cinematography is central to *Sicario*'s articulation of border liminality. Deakins repeatedly uses occluded frames of car windows, windshields, rear-view mirrors, night-vision goggles, and thermal-imaging scopes to limit both characters' and viewers' vision. This visual restriction is not incidental: it constructs a cinematic world in which knowledge is always partial, where danger emerges from the periphery rather than the event itself.

Jóhann Jóhannsson's score shapes the film's borderland atmosphere with an auditory architecture of dread. The low, pulsing bass motifs mimic the hum of helicopters, the thrum of engines, and the tremor of distant explosions. These sonic elements form a soundscape of anticipation, suggesting violence long before it appears on screen. Diegetic sound further heightens liminality; wind moving across open desert, the muffled interior acoustics of armoured vehicles, the sudden silence before an ambush, and the indistinct rumble of border crossings. The film's sound design acts as an affective guide through landscapes where the enemy is never clearly located. Much like *Threads* uses silence to convey infrastructural collapse and *Beasts* uses storm acoustics to signal ecological upheaval, *Sicario* uses sonic minimalism to represent the border as a space of perpetual imminence. The film's temporal structure is defined by waiting, stalling, and anticipation. Time stretches in convoys creeping through traffic; in tense silences during border checks; in the unbearable lull before sniper fire. These moments produce ***durational liminality***, where characters exist in temporal suspension; neither in action nor in safety, but poised between the two. Meanwhile, the narrative returns repeatedly to border crossing as ritual: the same roads, the same checkpoints, the same bottlenecks. These repetitions create a sense of circular time, echoing the perpetual, unresolved dynamics of border enforcement itself.

Sicario occupies a hybrid space between thriller, war film, and procedural drama. This genre-blending reinforces the sense of border landscape instability. There is no single narrative code through which the frontier landscapes can be understood. Instead, genre mutates in tandem with landscape. This hybridity echoes the border's status as a zone where normative categories collapse. Just as sovereignty blurs, film form destabilises. *Sicario* is thus liminal not only in content but in cinematic structure. *Sicario* well visualises the U.S.–Mexico frontier as a threshold world, shaped by the overlapping forces of cartel violence, militarised policing, covert operations, and environmental exposure. The desert's bare expanses absorb bodies and secrets, urban spaces reveal their fragility under pressure, and tunnels dissolve the spatial demarcations of the nation-state. Hence border landscapes are not a stable line but a cinematic process; a movement across landscapes where danger is atmospheric, visibility partial, and time suspended. Through its precise formal strategies, Villeneuve constructs a landscape where sovereignty becomes nomadic, violence becomes ambient, and characters navigate spaces that are neither fully governed nor free from governance.

Cinematic War Ruins and Conflict Zones: Suspended Sovereignties, Haunted Landscapes, and the Temporalities of Aftermath

Conflict landscapes on screen are threshold environments where time stalls, bodies persist precariously, and social infrastructures lie in suspended animation. Conflict landscapes frame conflict not as a discrete historical event but as a durational condition, inscribed into architecture, atmosphere, and everyday gestures. Since the late twentieth century, the geopolitics of conflict have shifted from conventional interstate warfare toward civil wars, occupations, insurgencies, and frozen conflicts. The break-up of Yugoslavia, the division of Cyprus, the Korean armistice, prolonged occupations in Palestine, and post-9/11 militarised zones have produced persistent liminal territories: buffer zones patrolled but not governed, cities partially rebuilt yet visibly scarred, and demilitarised regions where sovereignty remains indefinitely deferred. Scholars refer to these zones as *imperial debris* (Stoler, 2013), spaces where ruin is not an endpoint but a continuing mode of habitation. International intervention, humanitarian governance, UN peacekeeping architectures, and unresolved territorial claims all contribute to landscapes defined by waiting, haunting, and political stasis. Cinema and television, in turn, represent these unresolved spaces not only through spectacular destruction but through atmospheres of exhaustion, slow reconstruction, and the lingering presence of violence embedded in walls, streets, and bodies.

Several major films depicting the Second World War construct war ruins as liminal terrains, spaces where the distinction between battlefield and graveyard collapses and where the physical environment becomes an active participant in the suspension of ordinary life. Terrence Malick's *The Thin Red Line* (1998) visualises the Pacific front as a lush yet wounded ecology: tall grasses, misty hills, and dense jungle form a paradoxical landscape of beauty and dread, one in which violence ruptures the organic continuity of the terrain. Malick's drifting camerawork and whispered voice-over transform Guadalcanal into a philosophical threshold where nature endures even as human sovereignty unravels. By contrast, *Enemy at the Gates* (2001) renders Stalingrad as a monumental ruin with collapsed factories, shattered railway lines, and skeletal apartment blocks filmed through dust, smoke, and winter light. Sniper warfare turns the ruined city into an intimate labyrinth, a space of hyper-attentive listening and confined visibility. The *mise-en-scène* of hollowed-out stairwells, jagged concrete, and rubble-filled streets conveys the halted temporality of siege warfare, where survival unfolds moment to moment, and the city's ruins function as both cemetery and fortress. *Come and See* (1985) pushes the representation of war ruins to their most harrowing extreme, visualising Belarusian villages as sites of absolute annihilation. Burnt-out farmhouses, blasted forests, and mass graves appear in long, unbroken takes that refuse narrative respite. The landscape becomes a traumatised entity, its devastation inscribed into the faces of survivors and the silence of abandoned homes. Across these WWII films, ruined environments operate as cinematic thresholds, capturing the moment when social, political, and ecological worlds collapse into suspended futurity.

Representations of contemporary Middle Eastern conflicts frequently frame occupied and bombarded cities as liminal geographies, where sovereignty is fragmented and daily life unfolds amid the remnants of war. Kathryn Bigelow's *The Hurt Locker* (2008) presents Baghdad not as a unified battlefield but as a city in perpetual aftermath: landscapes are fractured by blast craters, debris-strewn markets, shuttered storefronts, and half-destroyed rooftops. War here is ambient, distributed, and spatially unpredictable, the city becomes a living ruin, flickering between moments of stillness and sudden eruption. In *Paradise Now* (2005), Palestinian urban landscapes of narrow alleys, rooftop routes, checkpoints, and military barriers are rendered as everyday liminal architectures, where occupation produces perpetual uncertainty. Domestic interiors feel provisional; outdoor spaces carry the weight of surveillance and risk. The *mise-en-scène* of concrete walls, razor wire, and unfinished structures visualises a world where the present is politically foreclosed and the future unavailable. Similar spatial dynamics appear in Hany Abu-Assad's *Omar* (2013), where the separation wall and its surrounding landscape function as a literal and symbolic threshold dividing intimacy from danger, movement from containment. Ari Folman's *Waltz with Bashir* (2008) approaches Middle Eastern conflict through animation, blending memory, dream, and ruin. Beirut becomes a surreal, fluid warscape: nightmarish street corridors, burning cities, and floating bodies depict conflict as a psychic liminality, where traumatic visions and material destruction coexist. Across these films, the Middle East is represented not as a singular battlefield but as multi-sited threshold landscapes shaped by occupation, fragmentation, and the slow accumulation of ruin.

African war films often emphasise the interplay between movement, displacement, and ruined environments, visualising conflict as a spatial condition that transforms villages, cities, and forests into landscapes of loss and survival. *Hotel Rwanda* (2004) portrays the Rwanda capital Kigali during the genocide through its urban landscapes. Although violence is often offscreen, the architecture and atmosphere of the hotel and surrounding city landscapes articulate the liminality of coexisting hospitality and danger. In *Sometimes in April* (2005), post-genocide Rwanda is depicted through its devastated schools, deserted homes, and mass graves; ruins that operate as material testimonies to violence. Here, ruin is durational. The landscape holds trauma long after peace agreements are signed. Across these films, African conflict zones emerge as cinematic landscapes of liminal survival, where characters inhabit spaces that cannot return to their pre-war states yet have not transitioned into stable post-conflict futures. Cary Joji Fukunaga's *Beasts of No Nation* (2015) shifts the focus to rural warzones, using dense forests, abandoned villages, and makeshift rebel camps to construct a mobile conflict landscape. Long takes of marching fighters and handheld shots of chaotic skirmishes reveal a landscape where the boundaries between civilian and combatant, safety and threat, are continually erased. The jungle becomes a threshold terrain, both concealment and exposure, refuge and trap.

Post-9/11 war films and television depict conflict as a mobile, decentralised and durational phenomenon, emphasising landscapes where battles unfold without clear fronts or endpoints. HBO's *Generation Kill* (2008) exemplifies this through its convoy-based narrative: Humvees move through shifting terrains of deserts, palm groves, and abandoned towns creating a cinematic experience of being in transit through conflict. Wide desert frames evoke exposure and cramped interiors emphasise claustrophobic uncertainty. The convoy becomes a moving liminal zone, never arriving, never securing territory, always suspended between threat and waiting. *Zero Dark Thirty* (2012) visualises geopolitics as incomplete sovereignty with U.S. forces operating within foreign landscapes under ambiguous authority. Night-vision sequences, helicopter insertions, and satellite imagery create a world of partial visibility and remote engagement, positioning the counterterrorism landscape as a juridical grey zone. *Lone Survivor* (2013) renders the Afghan mountains as a labyrinth of cliffs and scrub where movement is perilous and sovereignty incoherent. Across these works, the post-9/11 battlefield is not a bounded site but an evolving network of threshold landscapes where time, control, and identity remain in flux.

From Liminal Threshold Worlds to Wonderlands

Across this chapter, liminality has emerged as a spatial, temporal, and affective condition through which landscapes disclose their deepest instabilities and transformative potentials. Whether shaped by radiation, climate collapse, hydraulic engineering, militarised borders, or the ruins of war, these terrains reveal how power, memory, and vulnerability are sedimented into the environments we inhabit and imagine. Cinema amplifies these dynamics, transforming real-world thresholds into intensified aesthetic experiences that allow viewers to sense ecological precarity, political suspension, and the unsettling openness of worlds in flux. Yet if liminal landscapes foreground crisis, uncertainty, and becoming, they also prepare the conceptual ground for the landscapes that follow. Chapter 3 turns from sites of suspension to sites of wonder, to wonderlands that hold enchantment, estrangement, and imaginative possibility. These are landscapes that do not simply mark thresholds of danger or dissolution but open onto alternate modes of perception, curiosity, and world-making. Having explored how landscapes fracture, destabilise, and unsettle, the next chapter asks how they might also astonish, delight, and re-enchant: how wonder itself operates as a spatial logic, an affective mode, and a cultural force within the broader ecology of cinematic and cultural landscapes.

References

Agamben, G. (1998). *Homo sacer: Sovereign power and bare life*. Stanford University Press.

Agier, M. (2016). *Borderlands: Towards an anthropology of the cosmopolitan condition*. Polity Press.

Ahmed, S. (2004). Affective economies. *Social Text, 22*(2), 117–139. https://doi.org/10.1215/01642472-22-2_79-117

Ahmed, S. (2006). *Queer phenomenology: Orientations, objects, others*. Duke University Press.

Anzaldúa, G. (1987). *Borderlands/La frontera: The new mestiza*. Aunt Lute Books.

Baviskar, A. (1995). *In the belly of the river: Tribal conflicts over development in the Narmada Valley*. Oxford University Press.

Bhabha, H. K. (1994). *The location of culture*. Routledge.

Bruno, G. (2002). *Atlas of emotion: Journeys in art, architecture, and film*. Verso.

Butler, J. (1990). *Gender trouble: Feminism and the subversion of identity*. Routledge.

Choy, T. (2011). *Ecologies of comparison: An ethnography of endangerment in Hong Kong*. Duke University Press.

Clark, N. (2011). *Inhuman nature: Sociable life on a dynamic planet*. Sage.

Coghlan, J., Minns, J., & Wells, A. (2005). *Seeking refuge: Asylum seekers and politics in a Globalising world*. University of Wollongong Press.

Cruikshank, J. (2005). *Do glaciers listen? Local knowledge, colonial encounters, and social imagination*. UBC Press.

De Genova, N. P. (2002). Migrant "illegality" and deportability in everyday life. *Annual Review of Anthropology, 31*, 419–447. https://doi.org/10.1146/annurev.anthro.31.040402.085432
De Genova, N. P. (Ed.). (2010). *The deportation regime: Sovereignty, space, and the freedom of movement*. Duke University Press.
De León, J. (2015). *The land of open graves: Living and dying on the Migrant Trail*. University of California Press.
DeLoughrey, E. (2019). *Allegories of the Anthropocene*. Duke University Press. https://doi.org/10.1215/9781478005582
Elsaesser, T. (2005). *European cinema: Face to face with Hollywood*. Amsterdam University Press.
Fassin, D. (2013). *Humanitarian reason: A moral history of the present*. University of California Press.
Flynn, M. (2021). *Fear and the Anthropocene: Chernobyl and Fukushima in the visual culture of nuclear disaster*. Routledge.
Gordillo, G. (2014). *Rubble: The afterlife of destruction*. Duke University Press.
Gordon, A. F. (1997). *Ghostly matters: Haunting and the sociological imagination*. University of Minnesota Press.
Gordon, A. F. (2008). *Ghostly matters: Haunting and the sociological imagination* (2nd ed.). University of Minnesota Press.
Harper, G., & Rayner, J. (Eds.). (2010). *Cinema and landscape*. Intellect.
Knutson, T. R., et al. (2010). Tropical cyclones and climate change. *Nature Geoscience, 3*, 157–163. https://doi.org/10.1038/ngeo779
Kwon, H. (2010). *The other cold war*. Columbia University Press.
Lefebvre, H. (1991). *The production of space*. Blackwell.
LeMenager, S. (2014). *Living oil: Petroleum culture in the American century*. Oxford University Press.
Lukinbeal, C. (2005). Cinematic landscapes. *Journal of Cultural Geography, 23*(1), 3–22. https://doi.org/10.1080/08873630509478229
Macdonald, S. (2013). *American ecocinema*. University of California Press.
Magee, D. (2011). The three gorges dam and China's energy dilemma. *Energy Policy, 39*(5), 2533–2541. https://doi.org/10.1016/j.enpol.2011.02.009
Maldonado, J. K., et al. (2013). The impact of climate change on tribal communities. *Climatic Change, 120*, 601–614. https://doi.org/10.1007/s10584-013-0746-1
Mares, P. (2011). *Borderline: Australia's treatment of refugees and asylum seekers*. UNSW Press.
Masco, J. (2006). *The nuclear borderlands: The manhattan project in post–cold war New Mexico*. Princeton University Press.
Mitchell, T. (2002). *Rule of experts: Egypt, techno-politics, modernity*. University of California Press.

Molle, F., Mollinga, P. P., & Wester, P. (2009). Hydraulic bureaucracies and the hydraulic mission. *Water Alternatives, 2*(3), 328–349. https://www.water-alternatives.org/index.php/alldoc/articles/vol2/v2issue3/73-a2-3-1/file
Mora, C., et al. (2013). The projected timing of climate departure from recent variability. *Nature, 502*, 183–187. https://doi.org/10.1038/nature12540
Morris-Suzuki, T. (2014). Touching the grass: Science, uncertainty and everyday life from Chernobyl to Fukushima. *Science, Technology & Society, 19*(3), 331–362. https://doi.org/10.1177/0971721814544423
Mountz, A. (2011). The enforcement archipelago: Detention, haunting, and asylum on islands. *Political Geography, 30*(3), 118–128. https://doi.org/10.1016/j.polgeo.2011.01.005
Navaro-Yashin, Y. (2012). *The make-believe space: Affective geography in a postwar polity*. Duke University Press.
Neimanis, A., & Loewen Walker, R. (2014). Weathering: Climate change and the "thick time" of transcorporeality. *Hypatia, 29*(3), 558–575. https://doi.org/10.1111/hypa.12064
Nixon, R. (2011). *Slow violence and the environmentalism of the poor*. Harvard University Press.
Peters, K., & Steinberg, P. (2015). The ocean in excess: Towards a more-than-wet ontology. *Dialogues in Human Geography, 5*(3), 247–264. https://doi.org/10.1177/2043820615609999
Povinelli, E. A. (2011). *Economies of abandonment: Social belonging and endurance in late liberalism*. Duke University Press.
Richardson, B. J. (2025). Cinematic activism and Tasmania's Lake Pedder: How film shaped environmental history. *Australian Historical Studies, 56*(1), 103–125. https://doi.org/10.1080/1031461X.2024.2373810
Rose, D. B. (2004). *Reports from a wild country: Ethics for decolonisation*. UNSW Press.
Sassen, S. (2014). *Expulsions: Brutality and complexity in the global economy*. Harvard University Press.
Scheele, J. (2012). *Smugglers and saints of the Sahara: Regional connectivity in the twentieth century*. Cambridge University Press.
Schwenkel, C. (2009). *The American war in contemporary Vietnam: Transnational remembrance and representation*. Indiana University Press.
Scudder, T. (2005). *The future of large dams: Dealing with social, environmental, institutional and political costs*. Earthscan.
Sharpe, C. (2016). *In the wake: On blackness and being*. Duke University Press.
Sobchack, V. (1992). *The address of the eye: A phenomenology of film experience*. Princeton University Press.
Steffen, W., Richardson, K., Rockström, J., Cornell, S. E., Fetzer, I., Bennett, E. M., Biggs, R., Carpenter, S. R., De Vries, W., De Wit, C. A., & Folke, C. (2015). Planetary boundaries: Guiding human development on a changing

planet. *Science, 347*(6223), 1259855. https://doi.org/10.1126/science.1259855

Stoler, A. L. (Ed.). (2013). *Imperial debris: On ruins and ruination*. Duke University Press.

Turner, V. (1967). *The Forest of symbols: Aspects of Ndembu ritual*. Cornell University Press.

Turner, V. (1969). *The ritual process: Structure and anti-structure*. Aldine.

Turner, V. (1974). *Dramas, fields, and metaphors: Symbolic action in human society*. Cornell University Press.

van Gennep, A. (1909). *The rites of passage*. Émile Nourry. (English edn commonly cited Chicago, IL: University of Chicago Press).

Weizman, E. (2011). *The least of all possible evils: Humanitarian violence from Arendt to Gaza*. Verso.

World Bank. (2000). *Dams and development: A new framework for decision-making*. World Bank. https://documents1.worldbank.org/curated/en/123081468765309134/pdf/multi-page.pdf

CHAPTER 3

Through the Looking Glass: Wonderland Landscapes, Uncanny Encounters, and Hyperreal Terrains

Abstract Wonderland landscapes are hyperreal and uncanny terrains that reorganise perception, destabilise ordinary assumptions, and teach new forms of spatial literacy. Drawing on Freud's uncanny and Baudrillard's simulation, the analysis argues that wonderlands are not escapist detours but cultural technologies that train subjects to live with uncertainty, attraction, and altered rules of reality. These landscapes blur reality and fantasy, authenticity and artifice, comfort and unease, revealing how "the real" is continually produced through fantasy structures, designed experiences, and symbolic systems. Wonderlands are defined less by chaos than by coherent worlds with unfamiliar rules. Across literature, cinema, games, and built environments, wonderland spaces appear confusing at first but become intelligible once their internal logics are learned. Entry into these worlds occurs through rupture, erosion, bodily transformation, or temporal displacement, after which recognition gradually replaces confusion. What initially seems impossible begins to make sense as repeated patterns, constraints, and consequences emerge. Unlike surrealism, which resists coherence, wonderlands reward attention, inference, and adaptation. They require subjects to learn how space, time, bodies, and causality operate differently, turning disorientation into a process of understanding rather than a dead end. In their representations, wonderlands function as laboratories of cognition and embodiment.

J. Coghlan and H. Nolan, *Landscapes That Speak*,
https://doi.org/10.1007/978-3-032-21757-8_3

Literary wonderlands construct coherence through conceptual consistency rather than sensory detail; cinematic wonderlands teach viewers how to read worlds through colour, rhythm, framing, and sound; interactive wonderlands demand bodily learning through movement, failure, and recalibration. The analysis also foregrounds the politics of wonder: enchantment is routinely tied to regimes of power, including colonial fantasy, gendered and ableist norms, and the concealment of labour and environmental cost beneath spectacle. Finally, Indigenous and non-Western counter-wonderlands challenge Euro-American assumptions by presenting worlds that are not broken or unreal, but organised through alternative cosmologies in which land, memory, and nonhuman agency are already relational.

Keywords Uncanny · Hyperreality · Simulation · Fantasy · Perception · Immersion · Worldbuilding · Cosmology

Wonderland landscapes create worlds where the usual rules no longer apply. They invite us into spaces that feel surreal, uncanny, or heightened beyond everyday reality, and in doing so they train us to move through uncertainty, attraction, and shifting ways of being. These landscapes do not simply offer escape; they rehearse how we learn to live with instability, curiosity, and the possibility that the world might be organised differently. Drawing on Freud's idea of the *uncanny* (Freud, 1919) and Baudrillard's analysis of *hyperreality* (Baudrillard, 1994), these spaces unsettle the familiar binaries of reality/fantasy, authenticity/artifice, and comfort/estrangement (Eco, 1986; Žižek, 1991). But wonderlands are not trivial escapism. It is a cultural form that actively disrupts how people perceive place, identity, and experience, revealing how easily the boundaries of the real can shift. Building on Baudrillard's work on *simulation* and *simulacra* (1994), Vidler's *architectural uncanny* (1992), and Caillois's reflections on mimicry and spatial play (1961), wonderland landscapes emerge as sites where enchantment and anxiety, desire and disorientation coexist. These environments work through immersive, multisensory design, provoking emotional complexity, challenging perception and reshaping how social identities and cultural expectations are formed (Massumi, 2002; Thrift, 2008; Urry & Larsen, 2011). Wonderlands seek to unsettle the line between lived reality and imagined worlds, drawing attention to the ways in which landscape as experience is always mediated by cultural fantasies (Pine & Gilmore, 1999).

Across literature, film, games, and art, wonderlands disrupt what Husserl called the *natural attitude* of everyday perception; the assumption that the world will behave as expected (Husserl, 1970). These disruptions reveal alternative rules, grammars, and symbolic orders that the viewer or participant must learn before the world becomes intelligible. Wonderland narratives use disorientation to sharpen perception, turning confusion into a pathway toward understanding. They function as pedagogical spaces: laboratories in which readers, viewers, and players learn how new worlds hold together, and how coherence can be built from unfamiliar starting points. Whether experienced through the conceptual impossibilities of fiction, the visual distortions of art, the sensory demands of cinema, or the embodied challenges of gameplay, wonderlands teach forms of literacy that extend beyond fantasy. They model how subjects adapt to landscape instability, navigate unfamiliar spaces, and cultivate critical awareness in the face of the unexpected. Žižek's (1997) argument that fantasy is not an escape from reality but the structure that makes reality feel coherent clarifies this dynamic. Fantasy provides the interpretive framework through which desire is learned and objects become meaningful. A wonderland forms when moments of strangeness, talking animals, impossible physics, and unfamiliar environments and landscapes begin to reveal a pattern. Anomalies become signs of an underlying order. Recognition transforms estrangement into knowledge. This process hinges on how the subject responds to the limits of understanding. Adorno warns that modern reason often reacts defensively to strangeness, forcing new experiences into pre-existing categories and reducing difference to sameness, a tendency he called *identitarian thinking* (Adorno, 1973). Cassirer, by contrast, argues that meaning expands when new symbolic forms arise. Instead of flattening the unfamiliar, these forms create new ways of organising experience (Cassirer, 1944). A wonderland sits precisely at the tension between these two impulses.

Wonderland Landscapes: Strangeness and Grammar

Wonderland landscapes emerge not from pure absurdity but from a shift in relational perspective, a misalignment between the observer's expectations and the world's internal logic. In *Alice's Adventures in Wonderland* (1865), the Queen, the Cheshire Cat, and the card-soldiers inhabit a coherent order; it is Alice's inherited assumptions that render the world strange. This asymmetry reveals the foundational structure of wonderlands: disorientation arises when two systems of sense meet, and only one recognises the other. Urry and Larsen's *The Tourist Gaze 3.0* (2011) makes this dynamic visible in everyday experience, showing that perception is shaped not by neutral vision but by cultural, historical, and material position. The *tourist gaze* clarifies how wonderland landscapes are produced and consumed as designed experiences rather than encountered places. Wonderland landscapes operate through a learned way of seeing: they train viewers and visitors to recognise certain environments as magical, extraordinary, or immersive, even when those spaces are highly managed, staged, or technologically mediated.

Like the *tourist gaze*, wonderland landscapes depend on repetition, visual saturation, and expectation. They invite awe and enchantment while quietly disciplining perception, guiding where attention should linger, what should be photographed, and how pleasure and belonging are meant to feel. In this sense, wonderland landscapes do not suspend reality so much as reorganise it, turning space into spectacle and encouraging forms of ontological play that remain tightly structured by cultural and economic logics. As perspectives shift, estrangement becomes legible without collapsing into sameness. In this way, wonderlands rehearse the skills of critical literacy, learning how to inhabit unfamiliar orders while preserving the integrity of difference. The moment of entry formalises this shift. Rupture throws subjects into new laws instantly, as when Alice tumbles through the rabbit hole into altered spatial and bodily rules. Erosion describes a slower dissolution of categories, seen in Brandon Sanderson's *Stormlight Archive* (2010–), where the laws of nature subtly unravel. Bestowal reconfigures the perceiving body itself, as in *Avatar* (2009), where Jake Sully's mind is transplanted into a Na'vi body. Temporal dislocation, exemplified by *Outer Wilds* (2019), forces recognition to unfold across repeated 22-minute loops in which memory persists but the world resets. Across these modes, disorientation is not failure but method, the structural precondition through which recognition must occur.

Whether recognition arises depends on the willingness to confront anomaly. In *The Matrix* (1999), most people dismiss glitches as perceptual error, maintaining illusion through denial. Cypher recognises the system's artificiality yet rejects the labour of awareness, preferring the comfort of fabricated order. Thomas Kuhn's *Structure of Scientific Revolutions* (1962) describes this threshold more broadly: anomalies only provoke transformation when acknowledged as structural rather than accidental. Wonderland exists, therefore, only for those who register strangeness as meaningful; for those inside its logic, the world appears natural. Recognition is never instantaneous, it unfolds through the temporal demands of each medium. Literature trains inference: Carroll's bodily distortions in *Alice* (1865) and Jorge Luis Borges's infinite architectures in "The Library of Babel" (1941) teach readers to assemble coherence from scattered clues. Film cultivates sensory patterning: the threshold grammars of *Pan's Labyrinth* (2006) or the recursive dream-logic of *Mulholland Drive* (2001) guide viewers toward hidden structures. Games require enactment: meaning emerges through repeated failure and adaptation in *Dark Souls* (2011) or through recursive time in *Outer Wilds* (2019). Across forms, wonderlands convert temporal confusion into structured learning.

This systematic estrangement distinguishes wonderlands from surrealism. Surrealist cinema deliberately resists coherence, refusing to let estrangement crystallise into a learnable order. In *Un Chien Andalou* (1929), a razor slicing an eye or ants pouring from a hand bear no causal relation; meaning is governed by unconscious association rather than rule. *Eraserhead* (1977) extends this disorientation into a post-industrial nightmare, where sound and gesture loop without stabilising into pattern. Surrealism offers no grammar to learn, no structure to inhabit; it resists the very possibility of recognition. Wonderlands, by contrast, embed estrangement within systems that can be inferred, tested, and mastered. Their strangeness is patterned, not arbitrary. Žižek's claim that fantasy structures reality (1997) illuminates why wonderlands function pedagogically. Fantasy provides scaffolding through which subjects make sense of desire, logic, and order. Wonderlands expose this scaffolding by rearranging rules that usually remain unseen. Classification becomes a primary tool of this exposure. In *Annihilation* (2018), the Shimmer refracts light, memory, and biology, with refraction becoming ontology, producing hybrid organisms that materialise the grammar of transformation. Gestalt psychology explains the viewer's ability to navigate this world: perception naturally organises fragments into coherent wholes. Wonderland exploits this perceptual plasticity, forcing the mind to reorganise itself. Temporal anticipation deepens this pedagogy. Bloch's *Not-Yet-Conscious* (1986) shows that perception is always oriented toward futures still forming.

In *Annihilation*, mutations point toward a synthesis not yet realised. Learning becomes predictive rather than retrospective. Wonderlands thus operate as laboratories of cognition: estrangement reveals hidden rules, compelling subjects to build new interpretive frameworks. Confusion is not an impediment but the raw material of literacy.

Environmental grammar is the mechanism through which wonderlands become intelligible. Once recognition reveals that anomalies recur with pattern rather than randomness, grammar specifies how these patterns operate. Grammar links action to consequence, transforming landscapes from passive scenery into active pedagogical agents. Games demonstrate this especially well. In the 2D action-adventure game, *Hollow Knight* (2017) biomes adhere to distinct ecological and behavioural rules: fungal spores drift differently in the Fungal Wastes, acid reshapes movement in Greenpath, and Deepnest teaches vigilance by withholding visual cues altogether. Environmental sounds like skittering in tunnels, echoing footsteps, or distant screeches signal danger long before it is seen, turning auditory pattern into a learning system. *Subnautica* (2014) similarly trains systemic anticipation: depth alters pressure, bioluminescent flora mark safe zones, and the behaviour of leviathans teaches players to read the ocean's rhythm as method, not threat.

Cinematic wonderlands construct perceptual grammar rather than embodied one. In *The City of Lost Children* (1995), recurring motifs such as fog, winding alleys, amplified sound and distorted lensing train viewers to interpret the city landscape's dreamlike logic as rule rather than ambience. Terry Gilliam's *The Adventures of Baron Munchausen* (1988) teaches through tonal shifts: each realm (the moon, the underworld, the battlefield) obeys a separate aesthetic logic that viewers must learn to decode. *Coraline* (2009) uses repeated colour signatures, spatial distortions, and mirror-world symmetries to instruct audiences in reading the Other World's gradually tightening rules. Literary wonderlands construct grammar through conceptual impossibility. In Mervyn Peake's *Gormenghast* (1946–50), the castle's endless corridors, ritualised schedules, and architectural excess form a grammar of constraint: characters move within a world governed by ceremonial logic rather than physical law. In Stanisław Lem's *The Cyberiad* (1965), mathematical jokes, logical paradoxes, and engineered creatures collectively train the reader to expect that every invented impossibility follows a rule-based, algorithmic coherence. In Margaret Atwood's *The Heart Goes Last* (2015) alternating utopian and carceral environments reveal a grammar of behavioural conditioning rather than spatial fantasy teaching readers that wonderlands need not be supernatural to require new forms of sense-making.

Literary Wonderlands: Conceptual Coherence Without Sensory Support

Literary wonderlands differ from cinematic or interactive ones in a crucial way: they must construct coherence without the aid of sensory scaffolding. Because nothing is visibly shown or sonically rendered, readers must assemble the world from linguistic cues alone, building impossible or contradictory landscapes through imagination, inference, and conceptual labour. This makes literature uniquely suited to generating wonderlands whose rules may violate physical law yet remain internally consistent; worlds that can be *thought* but not *seen*. H. P. Lovecraft's R'lyeh is the fictional, submerged, non-Euclidean city introduced in the short story "The Call of Cthulhu" (1928). It is the place where the Great Old One, Cthulhu, lies in a death-like sleep, "dead but dreaming." R'lyeh sits deep in the Pacific Ocean, emerging only when cosmic alignments temporarily shift. Lovecraft describes it as a city built from impossible geometries; angles that behave as if they were both acute and obtuse, planes that tilt against rational orientation, and architecture that induces vertigo because it contradicts the perceptual expectations of the human mind. Its non-Euclidean geometry cannot be pictured, because the angles behave in ways impossible within human perception; the city folds space according to principles the mind struggles to grasp. Yet the descriptions are systematic enough that a conceptual grammar emerges, an alien geometry the reader learns to navigate despite its resistance to visualisation. Lovecraft's horror works not through sensory shock but through the disciplined effort required to imagine an order that exceeds embodiment.

Jorge Luis Borges develops a similar conceptual wonderland in "The Library of Babel" (1941), where an infinite library composed of identical hexagonal rooms contains every book that could ever be written. Its vastness cannot be pictured in any literal sense; readers can only grasp it through the combinatorial logic that makes such infinity possible. Borges replaces sensory detail with structural principle: the rules governing how chambers connect, how books permute through every arrangement of letters, and how meaning becomes indistinguishable from noise when all possibilities exist at once. The story transforms abstraction into architecture, compelling readers to navigate landscapes where order emerges through mathematical possibility rather than material description. In this way, Borges pushes the wonderland form toward pure conceptualisation, demanding that coherence be inferred from logic rather than imagined through sight. Contemporary fantasy extends this literary grammar of impossibility. In Brandon Sanderson's wider Cosmere universe, an interconnected sequence of novels beginning with *Elantris* (2005) and extending through series such as *Mistborn* (2006–) and *The Stormlight Archive* (2010–), worlds are bound together by a shared metaphysical architecture, requiring readers to navigate rule-based systems that span planets, histories, landscapes and ontological realms. All are bound by shared metaphysical laws across physical, cognitive, and spiritual realms. Readers learn the world not through sensory immersion but by grasping relations between oaths, energies, metaphysical forces, and their consequences across worlds, environments and landscapes. The conceptual coherence of the Cosmere depends entirely on rule inference: readers assemble the world by deducing the underlying laws that govern its impossible phenomena.

Other writers exploit similar strategies. China Miéville's *The City & the City* (2009) constructs an urban wonderland where two cities occupy the same physical landscapes, but citizens must unsee one another. The spatial logic is impossible to visualise, but perfectly coherent conceptually; grammar replaces geography. Gene Wolfe's *The Book of the New Sun* (1980–83) likewise builds a world where time, memory, and technology intertwine in ways that can be understood only through patient abstraction. In both cases, literature teaches readers to navigate wonderlands by conceptualising systems that cannot be sensorially apprehended. These literary wonderlands reveal the distinctive power of language: by withholding sensory certainty, literature obliges readers to treat estrangement as a catalyst for conceptual reasoning. Impossible cities, infinite libraries, overlapping realities, multi-realm cosmologies and unhinged landscapes become coherent only when readers learn their rules. In this way, literary wonderlands enact the most abstract form of wonderland landscapes cultivating a capacity to think across contradiction, to sustain alternative logics, and to build ordered worlds entirely within the mind.

Cinematic Wonderlands: Sensory Grammar

Cinematic wonderlands operate by remaking the conditions under which perception becomes intelligible. They do not simply depict impossible worlds; they construct *sensory logics* that teach viewers how to inhabit estrangement. Unlike literary wonderlands, cinematic wonderlands reorganises perception through audiovisual patterning. Its wonderlands are not anchored in narrative exposition but in rhythmic recurrence of textures, colours, sounds, and camera behaviours that gradually reveal an underlying system. To enter a cinematic wonderland is to submit to a perceptual retraining: coherence emerges not from explanation but from learning how to *see* and *hear* according to the world's own rules. *The Fountain* (2006) offers a clear expression of this logic by collapsing three timelines (Inquisition-era Spain, contemporary medicine, and far-future cosmic meditation) into a single sensory continuum. Darren Aronofsky' philosophical science-fiction film links these multiple timelines through symbolic landscapes. Jungles, bodies, and cosmic spaces function as metaphysical environments rather than mere settings, folding time, mortality, and devotion into one another. The film uses these wonderland landscapes to explore life, death, and transformation through a mode of ontological play rather than linear narrative. The film does not differentiate these worlds through narrative cues alone but through a disciplined grammar of light, motion, and chromatic saturation. Gold, applied as flare, halo, or cosmic dust, becomes the film's ontological stabiliser, signalling that each temporality is a facet of the same metaphysical continuity. Gradually, viewers learn to read light as discourse, with shifts in luminosity articulating emotional intensities, and recurring circular camera motions bind disparate worlds into a single perceptual field. Estrangement acquires structure because the film's sensory motifs recur with ritualistic precision. The wonderland coheres not despite its temporal fragmentation, but because its perceptual rules remain constant across time.

A very different sensory pedagogy unfolds in *MirrorMask* (2005), where Dave McKean's hybrid aesthetic (part live-action, part illustrated and part digital modelling) renders a world whose coherence depends on learning to see line, distortion, and gesture as structural cues. Buildings wobble; faces flatten and re-form; shadows behave as independent entities. But these distortions follow consistent rules. Angles buckle when emotional pressure mounts; scale shifts when Helena's self-perception falters. The film's wonderland teaches spectators to read symbolic instability as spatial logic: perspective becomes expressive rather than photographic. What first appears as stylistic eccentricity becomes, through repetition, a grammar that reveals how the world processes internal states through architectural and bodily deformation. In this sense, *MirrorMask* teaches viewers to interpret design as ontology. The film is defined by its surreal, illustrated landscapes, which operate as psychological and emotional terrain rather than coherent geography, staging a wonderland world in which identity, fear, and desire are externalised and positioning landscape as a space where inner transformation becomes visible.

Enter the Void (2009) radicalises this sensory training by severing perception from the human body altogether. Gaspar Noé's floating first-person camera hovering above streets, plunging through ceilings, sliding into memories forces viewers into a disembodied mode of seeing. The film's neon colour spectrum and deep bass pulses do not decorate its world; they constitute its ruleset. Colour surges signal transitions between life, death, memory, and hallucination. Sound locates the viewer when point-of-view cannot. Noé's Tokyo is a wonderland not because it is fantastical but because the perceptual apparatus itself becomes alien: the camera sees without a stable body, and viewers must develop new spatial orientations to follow it. Coherence emerges only when spectators learn to treat colour and drift as the primary coordinates of experience. *Enter the Void* links directly to wonderland landscapes by pushing their logic to an extreme. Rather than reshaping space, the film reshapes perception itself, turning the city into a world governed by unfamiliar sensory rules. Like a wonderland, Noé's Tokyo demands that viewers learn new ways of seeing and orienting, where colour, rhythm, and drift replace stable landscapes, and meaning emerges only through adaptation to an altered perceptual order.

This logic of patterned estrangement also governs *Paprika* (2006), whose dream-world unfolds through metamorphosis rather than stable form. Satoshi Kon's cuts behave as spatial mechanisms: a zoom becomes a passage between dreams, a match-cut becomes a physical transformation, and a change in colour temperature signals a shift in psychic territory. Crucially, Kon uses animation's plasticity to insist that continuity is metaphorical, not physical. The wonderland's rules become legible only when the viewer recognises that transitions, not objects, are the true units of meaning. Movement, rather than substance, structures this world. As motifs recur; a parade of consumer objects, melting cityscapes, shifting identities teaching viewers to track coherence not through linear causality but through rhythmic pattern. The wonderland is pedagogical because its impossibilities recur with formal consistency. The multiverse world of *Everything Everywhere All at Once* (2022) extends this sensory grammar into maximalist excess. Universes differentiate themselves through changes in aspect ratio, colour palette, editing tempo, even the choreography of bodies. At first, these shifts appear chaotic. But as the film progresses, repetition stabilises them into legible codes: a square aspect ratio with desaturated tones signals the bleak IRS universe; soft cinematic lighting and expressive close-ups signal the romantic movie-star timeline; frenetic montage and absurdist props signal juvenile multiverse branches. The viewer gradually learns to identify each universe through sensory cues alone. These patterns reveal that the multiverse, despite its chaos, operates through strict stylistic differentiation. It becomes a wonderland because estrangement is rule-governed, and recognition of these patterns becomes necessary for narrative and emotional comprehension.

Interactive Wonderlands: Embodied Literacy

Interactive wonderlands differ from literary, cinematic, and aesthetic forms in one decisive respect: they require the body to participate in the production of meaning. Where novels invite imaginative construction and film reorganises perception, interactive games force subjects to *act* within estranged landscapes, making coherence a practice rather than a premise. In these worlds, literacy does not emerge from observation alone but through repeated gestures of movement, failure, recalibration, and attunement. The player learns a wonderland through their body: its rhythms, hesitations, and habits gradually synchronise with the world's internal grammar. This is the distinctive epistemology of interactive media where knowledge is embodied and enacted rather than imagined.

Dark Souls (FromSoftware, 2011) offers one of the clearest formulations of this embodied pedagogy. Lordran, its labyrinthine world, is a spatial wonderland whose impossible landscapes loops back upon itself in unexpected and disorienting ways. Staircases fold into catacombs, battlements descend into swamps, and distant towers reveal themselves later as familiar thresholds viewed from altered perspectives. Its landscapes teach through loss. Death restarts the world but preserves experience, while each failure is not a setback but a refinement that reorders the player's bodily anticipation of danger. Vigilance becomes a learned reflex. Silence is no longer emptiness but an acoustic warning, a narrow doorway signals potential ambush and a clearing becomes a paradox either a gift and a threat. Here, wonderland grammar is encoded into posture, timing, and reflex. Understanding its landscapes is literally a matter of muscle memory. Where *Dark Souls* teaches vigilance, *Outer Wilds* (Mobius Digital, 2019) teaches players to inhabit temporality itself. In its 22-minute recursive loop, the solar system becomes an unfolding wonderland of cosmic logic. Planets collapse, sands shift between celestial bodies, oceans invert, and landscapes erodes in real time. None of this can be mastered through static observation. Knowledge accumulates through cycles of partial journeys, failed experiments, and fleeting discoveries carried forward into the next iteration. *Outer Wilds* teaches that understanding is an iterative dialogue between memory and recurrence, a wonderland where time is not a neutral container. Players learn to anticipate transformation like a comet's trajectory, a rising ocean, or a collapsing sun. Literacy becomes temporal attunement.

If *Outer Wilds* trains temporal literacy, *The Legend of Zelda: Breath of the Wild* trains ecological literacy. Hyrule is a systemic wonderland governed by weather, physics, chemistry, and the seasonal behaviour of plants and animals. Rain renders cliffs too slick to climb, lightning hunts metal, fire spreads with wind, and freezing temperatures sap vitality. Nothing is arbitrary. These systems function as environmental grammar that rewards curiosity and punishes inattention. To climb a mountain, players must first read the sky, to cross a field, they must read the grass, the wind, the clouds. Survival emerges from ecological attunement. *Breath of the Wild* turns the body into a barometer of an estranged but legible landscapes, where literacy requires not only recognition of cues but anticipation of cascading effects. Where Hyrule emphasises improvisational creativity, *Icarus* (RocketWerkz, 2021) imposes environmental severity. Its terraformed landscapes resemble Earth but harbour extinct megafauna, unpredictable storms, and fragile biomes that punish complacency. Icarus is a wonderland of environmental precarity: storms flatten structures, cold drains vitality within seconds, and oxygen scarcity transforms movement into a strategic gamble. Here, grammar is disciplinary rather than improvisational. Players learn to respect environmental rhythms not through curiosity but through necessity. A failure to prepare in the case of leaving a fire unattended or misreading the onset of a storm results in swift and irreversible loss. Icarus turns survival into pedagogy, embedding climate literacy into action and caution into muscle memory.

Minecraft (Mojang Studios, 2011) cultivates combinatory imagination. In its sandbox world, blocks of matter (wood, stone, iron, obsidian) are governed by a discreet grammar of transformation. Recipes are not provided; instead, the player learns through experimentation, pattern recognition, and abstraction. Coal plus stick becomes torch, grain aligned horizontally becomes bread and iron arranged in a square becomes a minecart. Each discovery reveals not an isolated fact but a principle of combination, teaching players to infer the logic of the world and its landscapes through the relational properties of its materials. *Minecraft* is a wonderland where the body learns by making, and where meaning arises from touch, placement, and transformation. This embodied literacy extends into analogue form through *Terraforming Mars* (FryxGames, 2016), which models planetary-scale landscape transformation as a strategic ecology. Though played on a tabletop rather than through an avatar, the game functions as an interactive wonderland: every tile placed, ocean raised, or temperature increased is an intervention into a systemic environment. The player learns to balance immediate gains with long-term planetary effects, coordinating resource scarcity, climate modification, and environmental interdependence. *Terraforming Mars* teaches that even abstract representations can operate as embodied pedagogy, where the hand placing tokens becomes a participant in ecological world-building. Interactive wonderlands achieve something distinct within landscape studies and narrative theory; they remake landscape as practice. Space becomes system, action becomes inquiry, and perception becomes literacy. Wonderlands demonstrate how estranged landscapes become legible only when inhabited, and how meaning develops through the body's negotiation with the world's internal logics. In these environments, knowledge is not simply interpreted, it is lived.

Ecological Wonderlands: Nature Made Legible

Ecological wonderlands occupy an uneasy space between revelation and distortion. They resemble fictional wonderlands in that they render the world newly visible, reorganising perception so that ecological processes appear coherent, patterned, and meaningful. Yet unlike fictional environments which are carefully engineered to produce navigable systems, natural ecologies are not designed to teach. Their complexity, indifference, and opacity far surpass the clean rule-sets of interactive or cinematic worlds. Ecological wonderlands arise only through mediation: when cameras, editing, narrative framing, and cultural expectation reorganise the natural world into forms that human perceptual and emotional apparatuses can readily grasp. They make the world legible, but only by filtering it through layers of aesthetic, technological, and ideological selection. This mediated legibility is most powerfully visible in the evolving grammar of David Attenborough's documentary corpus, from *Life on Earth* (1979) to *Our Planet* (2019). Technological advances of heli-mounted cameras, microdrones, ultra-slow motion, macro rigs, satellite imagery, night-vision, and CGI-supported reconstruction have reshaped what ecosystems look like and how they are understood. Aerial shots impose geometric order on migrations and river systems, time-lapse sequences condense days or seasons into minutes, revealing ecological rhythms ordinarily imperceptible and macro cinematography exposes life in scales beneath the threshold of naked vision. These techniques produce what Brockington (2009) has termed *spectacular environmentalism*, in which environments appear intelligible because technology has reorganised their temporal and spatial scales into coherent visual rhythms.

This intensifies in narrative-driven ecological wonderlands like *My Octopus Teacher* (2020). Although the film offers powerful glimpses into octopus cognition and kelp-forest ecology, its emotional architecture transforms the animal into a hyperreal protagonist. Through voice-over confession, intimate close-ups, and anthropomorphic framing, the octopus's camouflage, foraging, and brooding are rendered less as ecological strategies and more as emotional cues aligned with human drama. As Nolan's critique of the *hyperreal animal* (2025) argues, such framing produces an organism more symbolically compelling than its real-world counterpart, reciprocal, vulnerable, sacrificial. The octopus becomes a mirror rather than a subject; its ecological complexity collapsed into narrative grammar. The world is not encountered but translated into legible feeling. Even theoretically sophisticated frameworks struggle under this pressure. Donna Haraway's ethics of *becoming-with* (2008, 2016) attempts to develop multispecies accountability by grounding ethics in relational entanglement. This model often privileges visibility: to matter ethically, a being must appear, respond, or co-present itself in ways recognisable to human perception. Opaque, indifferent, microscopic, or geologic forms of life fall outside the field of relation. Despite its intentions, such ethics can reinscribe an anthropocentric demand that the nonhuman justify its value through recognisability rather than existence. This mode of affective translation reshapes landscape itself. The kelp forest becomes a scenic and emotional backdrop calibrated to human feeling rather than a complex ecological system with its own rhythms, scales, and indifferences. In this way, wonderland landscapes do not simply represent environments; they reorganise them into legible, narrativised spaces where ecological relations are subordinated to perceptual intimacy and emotional coherence.

Fiction films further demonstrate how ecological complexity becomes narratively reshaped. In *Okja* (2017), Bong Joon-ho translates industrial agriculture's systemic violence of deforestation, feedlots, genetic engineering, and methane emissions into a personalised rescue narrative. The hyperreal super-pig functions as a mnemonic device for global ecological harm, yet the film's emotional gravity centres on Mija's intimate bond with the super-pig. Structural environmental violence is reframed as the suffering of a single creature and critique collapses into a fable of loyalty and loss. In landscape terms, *Okja* transforms vast extractive environments into emotionally navigable terrain, converting forests, farms, and slaughterhouses into stages for individual attachment rather than systems of ecological entanglement, and in doing so renders structural environmental violence visible only through personalised loss. A different but related compression appears in *The Happening* (2008), which reduces plant communication, chemical ecology, and atmospheric toxicity to a moral allegory of environmental revenge. Complex vegetative processes (allelopathy, volatile organic signalling, adaptive stress responses) are distilled into intentionality, granting "nature" a grammar of warning and punishment tailored to human comprehension. Ecological wonderlands therefore illuminate and obscure in equal measure: they expand perceptual access to ecological processes while simultaneously simplifying, emotionalising, or anthropomorphising them. They are not ecologies, but mediations of ecology. Their power lies in the legibility they provide and their risk lies in the illusions of coherence they create.

Power, Politics, and Inequality in Cinematic Wonderland Landscapes

Cinematic wonderlands are often celebrated for their imaginative freedom, yet they are built from, and contribute to, highly structured *regimes of power* (Foucault, 1977, 1980, 1991). Their sensory enchantments operate alongside ideological training, teaching spectators how to navigate landscapes shaped by race, coloniality, gender, class, and labour. Here wonderlands reproduce privilege while disavowing the inequalities that sustain them meaning wonderlands are not politically neutral terrains of escapism but carefully structured landscapes where fantasies of possibility and exclusion are intertwined. Cinematic wonderlands often reproduce colonial aesthetics under the guise of mythic or speculative world-building. *Avatar* (2009), *John Carter* (2012), and *Valerian and the City of a Thousand Planets* (2017) deploy Indigenous-coded alien cultures, pristine ecologies, and frontier mythologies that legitimise white protagonists as explorers, intermediaries, or saviours. These films present planetary wonderlands cleansed of colonial history, re-staging conquest as a rite of empathy and self-discovery. Bell hooks's critique of white desire for the Other's spirituality (1992) resonates strongly here, as does Mbembe's theorisation of colonial spatial control (2003).

Even seemingly critical films sustain these logics. *Black Panther* (2018), while celebrated for its Afrofuturist aesthetic, imagines Wakanda as a fantastical African wonderland untouched by colonialism. The film stages a decolonial imaginary yet depends on the premise of an Africa that avoided historical violence thereby turning colonialism into an absence rather than a global structure, and replacing material history with aesthetic possibility. Similarly, *Beasts of the Southern Wild* (2012) creates a swampy wonderland (The Bathtub) that appears outside state power, yet it relies on racialised and poverty-coded bodies to produce a sublime landscape of resilience and wildness. Kathryn Yusoff's work on anti-Black geologies (2018) underscores how such films aestheticise disaster and precarity through the bodies of racialised subjects, rendering their hardship into scenic wonder.

Wonderlands frequently discipline gendered and sexual bodies into idealised cinematic forms. In *The Shape of Water* (2017), the fantasy of interspecies romance is framed through the hyper-feminised vulnerability of Elisa, whose difference (her muteness) becomes a narrative device to emphasise purity and emotional accessibility. Guillermo del

Toro's aquatic wonderland invites spectators to celebrate non-normative attachment while simultaneously reproducing gendered tropes of caregiving, sacrifice, and feminised empathy. Other films exert more overt disciplinary pressures. *Mad Max: Fury Road* (2015), despite its feminist reputation, constructs a desert wonderland in which the mobility and survival of women depend upon their status as symbolic reproducers (the Wives) or as warrior exceptions (Furiosa). The environment trains viewers to read female bodies either as commodities or saviours, sustaining the *orientation of desire*, a spatial organisation of bodies around heteronormative and masculinist imaginaries (Ahmed, 2006). Cinematic wonderlands also police bodily ability. *Doctor Strange* (2016), for example, frames mystical landscapes as accessible primarily to bodies capable of extraordinary physical discipline and visual-spatial mastery, aligning heroism with able-bodied virtuosity. Disabled characters appear only to be cured, surpassed, or morally corrupted, echoing Garland-Thomson's critique of the *normate template* (2011). In landscape terms, *Doctor Strange* configures its wonderlands as selective terrains that reward bodily conformity and mastery, turning space itself into a test of capacity. Mystical environments function as disciplinary landscapes, calibrated for bodies that can see, move, and endure in normatively heroic ways, while disability is positioned as an obstacle to spatial belonging rather than a legitimate mode of inhabiting the world.

The labour required to produce cinematic wonderlands is systematically concealed, both within the diegesis and behind the screen. In *Charlie and the Chocolate Factory* (2005), the Oompa Loompas constitute a racialised and miniaturised labour force whose work underpins the factory's fantastical abundance. Their replication through CGI erases their individuality and obscures the colonial template on which their servitude is modelled, what Harvey (2000) considers as the *aesthetic mask* that hides exploitative relations beneath spectacle. In *The Hunger Games* film series (2012–15), landscapes are not neutral settings but central mechanisms of power, control, and resistance. The dystopian nation of Panem is organised spatially through extreme inequality: the opulent Capitol, the regimented Districts, and the lethal arenas each embody distinct political and economic logics. The arena is the most explicit landscape of domination. Designed as a spectacular wilderness, it is a technologically engineered environment that turns forests, fields, deserts, and ruins into weapons. Weather, terrain, visibility, and scarcity are manipulated to discipline bodies, reward adaptability, and produce televised suffering. Nature here is not wild but curated, surveilled, and lethal, exposing how landscapes can be mobilised as instruments of authoritarian power. Beyond the arena, the District landscapes visualise structural violence. District 12's degraded mining environment of scarred land, toxic labour, and exhausted bodies contrasts sharply with the Capitol's excessive artificiality and abundance. These spatial contrasts naturalise class hierarchy, making inequality appear geographically inevitable. Over the series, however, landscapes also become sites of resistance: forests function as fugitive spaces, ruins enable insurgency, and the reclaimed terrain of District 13 reveals how alternative relations to land and survival can be organised. *The Hunger Games* frames landscape as a political actor: a system that disciplines, spectacularises, and punishes, but one that can also be reoccupied, repurposed, and turned against the regimes that designed it.

Cinematic wonderlands often reproduce class distinctions by framing subjective access to wonder as universal while encoding it as an elite privilege. In *Downton Abbey* (2019), the estate functions as a nostalgic wonderland for spectators, yet the film obscures class antagonism through sentimentalised depictions of servitude. The lush pastoral landscapes and choreographed interiors invite audiences to romanticise aristocratic order, turning class hierarchy into aesthetic pleasure. Similarly, the magical steampunk city of *Mortal Engines* (2018) presents mobility and abundance as aesthetic wonder while normalising the violent extraction and class warfare that sustain it. These films teach spectators to read spectacular wealth and mobility as naturalised features of wonderland rather than as class-specific affordances. Even darker cinematic wonderlands such as *Nightmare Alley* (2021) reproduce class anxieties by framing carnival spaces as illicit, uncanny landscapes from which upward mobility requires deception or moral compromise. Here, the wonderland becomes a classed trap: it teaches viewers to associate working-class spectacle with danger and elite spectacle with safety. Wonderlands teach audiences how to consume beauty, how to desire order, how to overlook labour, and how to imagine themselves as central to fantastical worlds. They quietly produce subjects aligned with dominant structures of whiteness, patriarchy, ability, consumption, and class aspiration.

Dystopian Wonderland

Dystopian wonderlands now sit at the centre of contemporary film and television because they merge enchantment with control. These landscapes combine the visual allure of fantasy with the coercive logic of dystopia, creating worlds that captivate even as they discipline. Rich colour, immersive atmosphere, and tightly organised space seduce viewers while quietly normalising surveillance, hierarchy, and unequal labour. Far from serving as neutral backdrops, dystopian wonderlands operate as ideological landscapes that shape how audiences learn to desire, adapt, and comply within systems of power. Their contradiction is central to their force: the very qualities that make these worlds pleasurable are also those that render domination beautiful, familiar, and difficult to resist. Clear expressions of the dystopian wonderland emerge in corporate-technological landscapes where design, aesthetic pleasure, and immersive architecture function as mechanisms of control. *Westworld* (2016–) is paradigmatic in this respect. Its frontier landscape is meticulously curated to produce enchantment, vast desert vistas, picturesque towns, and perfectly choreographed encounters, yet this beauty is inseparable from the corporate *necropolitics* governing the park. The wonderland's sensory richness conceals a system in which human guests enjoy frictionless freedom while the android hosts endure continuous cycles of violence, erasure, and exploitation. Wonder does not counteract domination; it enables it. The landscape's enchantment is precisely what permits the viewer, like the visitor, to overlook its brutal infrastructural logic. A similarly seductive corporate wonderland appears in *Severance* (2022–). Lumon Industries' interiors comprise a maze of pastel corridors, retro-futurist offices, and uncanny recreational spaces that together form an aesthetic regime of bureaucratic surrealism. The playful oddity of sensorially bland meeting spaces with manufactured sublime excesses disorients both characters and viewers, producing an affective instability that masks the violence of cognitive division and labour extraction. In this world, disorientation operates as a mode of discipline: the pleasurable strangeness of the environment distracts from the coercive system structuring employees' divided consciousness. The corporate workplace becomes a wonderland whose gentle absurdities obscure the machinery of control.

Other dystopian wonderlands arise from the logics of consumer capitalism, where wonder operates explicitly as seduction. In *Blade Runner*

2049 (2017), Los Angeles is reimagined as a neon-saturated, sensorially overwhelming cityscape built upon ecological collapse, racialised precarity, and the exploitation of replicant labour. The film's shimmering holograms, saturated colour fields, and monumental digital advertisements create a visually intoxicating environment that transforms dystopia into spectacle. In this sense, Villeneuve constructs what Yusoff (2018) considers a *racialised geologic dystopia*: a world whose aesthetic grandeur depends on the extraction and devaluation of specific bodies and environments. The wonderland is catastrophic precisely because it is beautiful. A similar dynamic structures *Minority Report* (2002), where the pleasures of frictionless mobility, personalised advertising, and translucent technocratic design conceal a system of pre-emptive policing and biopolitical categorisation. The film's interface aesthetics of gestural screens, luminous surfaces, and clean architectural lines produce a utopian sheen that hides the dystopian logic beneath it. Surveillance becomes palatable because it is wrapped in the promise of efficiency and visual delight. Wonder, in these landscapes, is capitalism's most effective camouflage. These films show how landscape itself becomes the primary vehicle through which capitalist power is normalised. Urban environments are designed as immersive wonderlands whose visual excess and sensory pleasure reorganise perception, training viewers to accept extraction, surveillance, and inequality as the background conditions of everyday life. In this way, dystopian landscapes do not simply represent systems of power; they aestheticise them, allowing domination to appear seamless, desirable, and spatially inevitable.

Urban dystopian wonderlands dramatise how hypermodern spectacle and systemic collapse coexist within the same spatial frame. *The Fifth Element* (1997) offers a vertical megacity of airborne taxis, neon advertisements, and stacked architectural layers that together produce an exuberant comedic wonderland. Yet this sensory excess, in line with Sontag's critique of how aesthetic intensity can overwhelm critical perception, both heightens affect and distracts from the authoritarian bureaucracies and systemic overcrowding that sustain the city's operation (Sontag, 1977). In *Ghost in the Shell* (1995, 2017), neon futurism and cybernetic augmentation construct a visually mesmerising metropolis in which bodies and landscapes blur. While the city dazzles with digital overlays and augmented bodies, it also normalises total surveillance, corporate militarism, and post-human alienation. Its wonder is inseparable from the dissolution of autonomy and the intensification of the networked subject.

Yet, Fritz Lang's *Metropolis* (1927) remains one of the clearest early exemplars of the dystopian wonderland. The upper city appears utopian with immaculate gardens, luminous towers, and choreographed leisure yet this visual paradise is sustained by the ceaseless labour of exploited workers relegated to subterranean machine halls. The city's beauty is legible only because its violence is hidden. This is the dystopian wonderland *avant la lettre*: a world where the enchantment of the few depends on the exhaustion of the many.

Beyond the Western Imaginary: Non-Western and Indigenous Counter-Wonderlands

Up to this point, we have examined wonderlands largely through Euro-American frameworks of Freud's uncanny, Baudrillard's hyperreality, Caillois's mimicry, and Žižek's fantasy structure, amongst others. Yet such definitions presuppose a particular metaphysics of land, world, and subject. Many Indigenous and non-Western traditions begin with fundamentally different ontologies: worlds in which spirits, ancestors, nonhuman agents, and cosmological forces already cohabit with the human; worlds in which permeability rather than separation structures place. The wonderlands that emerge from these cosmologies do not operate as distortions of reality but as continuations of enduring relational worlds. Rather than staging estrangement as a pedagogical shock, these counter-wonderlands foreground forms of knowledge that Western epistemologies have historically marginalised or suppressed. Acknowledging these alternative traditions is a critical reorientation of the concept of wonderland itself. It signals that Euro-American frameworks do not exhaust the possible grammars of strangeness or terrain of imaginative world-making. Indigenous and non-Western wonderlands expose how deeply the Western category of wonder is entwined with settler colonialism, secular rationality, and capitalist modernity, and how alternative cosmologies resist, reconfigure, or refuse these logics.

Indigenous futurisms offer one of the most decisive challenges to ideas of hegemonic wonderlands. Grace Dillon (2012) argues that Indigenous speculative forms construct futures not by rupturing from the past but by amplifying Indigenous cosmologies already grounded in reciprocity, spirituality, and land continuity. Kyle Whyte (2018) similarly notes that for many Indigenous peoples, apocalypse is not hypothetical; it has already occurred through colonisation. As such, Indigenous futurist wonderlands are not spaces of sudden estrangement but landscapes of ongoing survivance, worlds coherent to Indigenous protagonists yet rendered strange by colonial ignorance. *Night Raiders* (2021) exemplifies this logic. Set in a dystopian near-future where Indigenous children are seized by militarised state forces, the film stages a world that appears alienated or broken from a settler perspective but operates intelligibly through Indigenous protocols of kinship, resistance, and land-based knowledge. Estrangement arises not from fantastical impossibility but from the imposition of settler regimes that render Indigenous governance illegible. The wonderland here is a counter-landscape: a speculative world where Indigenous resurgence reasserts spatial orders that settler modernity attempted to erase. In Māori speculative writing, such as the stories collected in *Pūrākau*, wonderlands emerge from *whakapapa* (genealogical connection), *mana* (spiritual authority), and *mauri* (life force). These worlds do not require readers to learn unfamiliar rules, instead they reveal rules already in place but obscured by Western epistemologies. Wonderland becomes a form of epistemic return: the recognition that land has agency, ancestors shape place, and humans occupy one node in a wider cosmological network. In this sense, Indigenous counter-wonderlands invert the Western pedagogical model. They do not teach through confusion or surprise rather they teach through re-attunement, decolonial remembering, and relational accountability.

East Asian cultural forms provide another model of counter-wonderland, one rooted not in rupture but in porous cosmologies where the human and nonhuman worlds interpenetrate. Miyazaki's *Spirited Away* (2001) and *Princess Mononoke* (1997) articulate wonderlands structured through Shinto-inflected ecological logics where *kami* (spirits) inhabit rivers, forests, and abandoned infrastructures, thresholds between worlds are fluid and environmental imbalance manifests as spiritual disturbance abound. Estrangement here functions differently than in Western narratives, the viewer does not encounter a broken or uncanny world but one whose rules they have forgotten how to read. The bathhouse in *Spirited Away* is a particularly powerful example of environmental grammar. Its architecture teaches viewers that spaces have moods, histories, and spiritual resonances, that pollution is not symbolic but literal harm to spirits, and that labour, ritual, and reciprocity restore balance. The wonderland is thus a moral ecology, not a simulation. Recognition emerges through re-learning relations rather than discovering new rules. This is a pedagogical logic profoundly different from the Western *uncanny*. Chinese *xianxia* worlds, drawing on Daoist alchemy, *qi* cultivation, and cosmological balance, further complicate Euro-American understandings of wonder. These narratives, such as *The Untamed* (2019) or novels within the *Zhu Xian* and *Stellar Transformations* traditions stage impossibilities (flight, transformation, spiritual realms) that are entirely coherent within their metaphysical systems. The wonderland is not fantasy breaking realism but realism articulated through alternative metaphysics. Cultivation cycles, sect ecologies, and moral energetics form the grammar of *xianxia* worlds; their rules are internally consistent but externally incompatible with Western rationalism. Thus, estrangement arises from cultural translation rather than narrative rupture. These texts foreground landscape as a relational and ethical system rather than a spectacle of excess or control. Rivers, forests, and built environments operate as living terrains with memory, agency, and moral force, teaching viewers how to read space through reciprocity and balance rather than mastery. In this model, wonderland landscapes function not as escapes from reality but as invitations to recover alternative ways of inhabiting land, where estrangement arises from forgotten relations rather than broken worlds.

Afrofuturist cultural production reworks wonderlands through histories of racialisation, diaspora, and technological exclusion. Nnedi Okorafor's *Lagoon* (2014) transforms Lagos into a wonderland where alien arrival activates dormant ecological and spiritual rhythms. The estrangement is not the aliens but the intensified visibility of forces long present in the landscape: water spirits, linguistic multiplicity, infrastructural chaos, and futurist possibility nested inside postcolonial precarity. Similarly, Wanuri Kahiu's *Pumzi* (2009) imagines a post-apocalyptic East African future where water scarcity produces harsh technocratic environments. Yet the film's true wonderland is not the bunker-city but the desert outside it, where ecological rebirth defies technological determinism. Afrofuturist wonderlands, as Kodwo Eshun (2003) argues, are sites of speculative repair: landscapes where the future is not a break from the past but a means of rewriting traumatic histories and imagining new communal forms. Urban Afrofuturist spaces like Johannesburg's hybrid techno-spiritual geographies in *District 9* (2009) or the reimagined African metropolis of Wakanda in *Black Panther* (2018) further illustrate how wonderlands can function as critiques of colonial urbanism. Wakanda's vibranium-rich landscapes reconfigure futurity through African technological sovereignty rather than Western models of development.

The wonderland is thus political, it is a landscape structured by refusal of colonial extraction and assertion of Indigenous technological lineage.

The wonderland logic extends beyond media into real urban landscapes across East, Southeast, and South Asia. Tokyo's Shinjuku and Shibuya districts, Seoul's Gangnam, Shanghai's Pudong skyline, Jakarta's megamalls, and Singapore's hyper-regulated green futurism create sensory and architectural grammars that differ from Euro-American spectacles. These cities integrate digital layers, religious iconography, dense informal economies, and rapid infrastructural cycles in ways that feel disorienting to Western observers but are entirely legible to local inhabitants. For example, the Gardens by the Bay in Singapore function as a hybrid wonderland, part ecological demonstration, part climate-controlled fantasy, part technocratic display, yet grounded in Southeast Asian environmental planning rather than Western pastoralism. These landscapes reflect regional histories of tropical governance, colonial botany, and developmental-state futurism. They exemplify what Chua (2011) describe as *tropical modernism*, a spatial logic that blends cosmology, climate, and capital in ways irreducible to Western models of spectacle. These Indigenous and non-Western examples demonstrate that wonderlands are not universal structures of estrangement but culturally specific epistemologies of world-making. In Indigenous, Shinto, Daoist, and Afrofuturist cosmologies, wonder is not a break from the real but a form of continuous relation. The strange is not *uncanny* but familiar; the impossible is not irrational but metaphysically coherent. Wonderland landscapes in these traditions thus function as counter-theories; decolonial, relational, cosmological corrections to the Western imaginary.

Real-World Built Wonderlands: Theme Parks, Malls, Resorts, and Spectacular Cities

Real-world built wonderlands translate the logics of fictional landscapes into material form. As critics of themed and commercial space have shown, environments such as Disney parks, mega shopping malls, cruise ships, and master-planned resorts are designed as tightly scripted spatial systems that choreograph movement, affect, and behaviour (Sorkin, 1992). These sites rely on controlled circulation, immersive aesthetics, and the removal of friction to produce spaces that feel coherent, pleasurable, and inevitable, while concealing labour, exclusion, and infrastructural management behind seamless design. In this sense, built wonderlands exemplify what Baudrillard describes as the *hyperreal*: spaces where simulation does not mask reality but replaces it, becoming the dominant mode through which space is experienced (Baudrillard, 1983). As Urry and Larsen argue, such environments also train specific ways of seeing and inhabiting space, aligning pleasure with consumption and movement with surveillance (Urry & Larsen, 2011). Wonderland landscapes thus extend beyond screen and page into everyday life, operating as lived architectures that shape perception, desire, and compliance within late-capitalist spatial regimes.

Disney and Universal theme parks remain the paradigmatic examples of the wonderland as built landscape. Umberto Eco's visit to Disneyland led him to conclude that the park functions as a "fantastic past" that is "more real than real" (Eco, 1986, 43), a space that produces authenticity by fabricating it. Baudrillard similarly argued that Disneyland "exists in order to hide that it is the 'real' America" (1994, 12), an ideological screen that naturalises the very fantasies it stages. These parks operate through meticulously choreographed spatial grammars: sightlines that conceal backstage labour, forced perspective that manipulates scale, ambient soundtracks that regulate affect, and ride narratives that script emotional arcs. Each of these techniques functions pedagogically. Visitors learn to navigate the world through perceptual cues rather than explicit instruction, internalising a logic of movement, attention, and consumption that is both playful and disciplinary (Bryman, 2004). The built wonderland here is not simply an environment of attractions but a social architecture of behaviour. As Wasko (2001) argues, the park's promise of enchantment is inseparable from its regimes of order: surveillance systems, crowd management strategies, cleanliness protocols, and affective labour performed by 'cast members' produce a world in which fantasy and regulation coexist seamlessly. This echoes our chapter's broader claim that wonderlands teach new grammars of sense-making. In Disneyland, that grammar is aestheticised order; a world that trains the visitor to expect coherence, benevolence, and legibility, even as such coherence depends on the concealment of labour, cost, and environmental impact. The *uncanny* thus emerges not in the form of obvious strangeness but in the unsettling revelation that an apparently natural fantasy is sustained by invisible infrastructures of control.

If theme parks compress fantasy into walkable terrains, cities like Las Vegas and Dubai expand the logic of the wonderland into total urban ecologies. Michael Sorkin (1992, 217) describes Las Vegas as a "city of simulations," where signage, architecture, and entertainment fuse into an urban semiotic field that is both overwhelming and meticulously orchestrated. The Strip operates as a linear wonderland: a concatenation of themed environments; the Venetian, the Luxor, Paris Las Vegas that ask visitors to move through continuous cycles of estranged recognition. These sites enact what Hannigan (1998) names as the *fantasy city,* an urban landscape structured not around production or civic life but around spectacle, eventfulness, and perpetual consumption. Dubai extends this logic into a geopolitical key. As a city built on petro-capital, migrant labour, and hyper-accelerated construction, it materialises the wonderland as a global capitalist imaginary: artificial islands shaped like palm trees, indoor ski slopes in the desert, and climate-controlled megamalls that simulate winter beneath 45-degree heat. These landscapes behave like cinematic or game worlds, impossibly coherent, visually excessive, and structurally insulated from the environmental, social, and economic externalities that sustain them. Their uncanniness arises from their ambition to render the impossible normal.

Burj Khalifa is the world's tallest building and a defining feature of Dubai's spectacular urban landscape. More than an architectural feat, it functions as a vertical wonderland, translating desert modernisation, global capital, and technological mastery into a single iconic form. Rising from an arid environment dependent on imported labour, energy, and water, the tower aestheticises abundance and futurity while obscuring the extractive, ecological, and human infrastructures that sustain it. As landscape, the Burj Khalifa reorganises perception at a planetary scale: it makes vertical dominance feel natural, progress feel inevitable, and inequality disappear into spectacle. The World Islands is a vast artificial archipelago off the coast of Dubai, designed to replicate a map of the world through engineered land reclamation. Conceived as a luxury development, it transforms geography itself into spectacle, reducing continents to purchasable plots and converting planetary difference into private real estate. As a landscape, the project exemplifies hyperreal spatial logic: the 'world' is no longer encountered through distance, history, or relation, but miniaturised, aestheticised, and owned. Its ongoing instability of erosion, sinking, and stalled development reveals the tension at the heart of such wonderlands, where global ambition collides with ecological limits and the material resistance of land and sea. As the Burj Khalifa pierces atmospheric scale and The World Islands perform cartographic fantasy, Dubai demonstrates how *hyperreal landscapes* become political fantasies of mastery over climate, geography, and labour are not metaphors, but literal, built infrastructures.

Where theme parks and spectacular cities stage wonderlands overtly, malls and cruise ships embed wonderland logics within everyday spatial practice. The shopping mall, with its climate control, skylit atria, labyrinthine circulation paths, and absence of clocks, constructs what Crawford (1992) considers as *controlled spontaneity*. The mall simulates public space while enforcing private regulation; it mimics urbanity while circumscribing behaviour. Its promise of comfort, predictability, and endless choice echoes the affective logics of wonderlands: estrangement softened by familiarity, fantasy scaffolded by consumption. Cruise ships such as those analysed by Wood (2000) intensify this spatial fantasy by creating floating microcosms of plenitude and insulation. They resemble wonderlands not just in aesthetic form as spectacle, themed zones, and choreographed entertainment but in their environmental and logistical isolation from the real-world ecologies they traverse. Their self-contained infrastructures of power, sanitation, surveillance, and consumption reproduce the illusion of detached, frictionless mobility, a world where the sea becomes backdrop rather than environment. As with the cinematic wonderlands coherence emerges through rhythmic repetition: daily schedules, staged performances, architectural motifs. The ship becomes a pedagogical device, teaching subjects to inhabit a world where logistical dependency is masked as leisure autonomy.

What unites these diverse built environments is their function as ideological landscapes. They materialise the fantasy structures that Žižek (1997) argues organise the real, translating abstract desires of safety, abundance, mastery, and wonder into spatial form. Their hyperreal coherence is not accidental but engineered, teaching visitors what it feels like to move within a world where ambiguity is erased and contingency is controlled. These are not sites of escapism rather they are infrastructures through which subjects learn to inhabit late capitalist needs of perception, governance, and desire. Thus, built wonderlands close the loop between textual/virtual examples and the lived spaces of contemporary culture. They reveal that the wonderland is not a metaphor for estrangement but a material spatial regime that instructs, disciplines, and delights simultaneously. They demonstrate that landscapes of fantasy are landscapes of power, and that the hyperreal operates not only in fiction but across the architectural, economic, and climatic infrastructures of everyday life.

Wonderland landscapes reveal how patterned estrangement reshapes perception, reorganises meaning, and teaches subjects to inhabit unfamiliar orders with new forms of literacy. Across film, television, literature, games, and built environments, these spaces demonstrate that wonder is never merely decorative instead it is a spatial logic that unsettles habits of seeing, exposes the limits of rational coherence, and renders the familiar strange in ways that are both enchanting and disquieting. Yet if wonderlands foreground perception as instability, they also gesture toward deeper cultural structures that precede and exceed individual experience, structures that persist across time, ritual, belief, and collective memory. It is into this terrain that the next chapter turns. While wonderlands destabilise the real through estrangement and patterned deviation, mythical landscapes operate through continuity: they anchor worlds in cosmological order, cultural memory, and enduring symbolic meaning. Where wonderlands expose the fragility of the real, mythical landscapes reveal the narratives that hold worlds together. Chapter 4 explores how mythic terrains of sacred mountains, enchanted forests, origin sites, and cosmological thresholds function not as disruptions of reality but as its cultural foundations, shaping belonging, authority, and collective imagination across time.

References

Adorno, T. W. (1973). *Negative dialectics* (E. B. Ashton, Trans.). Routledge.

Ahmed, S. (2006). *Queer phenomenology: Orientations, objects, others*. Duke University Press.

Atwood, M. (2015). *The heart goes last*. McClelland & Stewart.

Baudrillard, J. (1983). *Simulations* (P. Foss, P. Patton & P. Beitchman, Trans.). Semiotext(e).

Baudrillard, J. (1994). *Simulacra and simulation* (S. F. Glaser, Trans.). University of Michigan Press.

Bloch, E. (1986). *The principle of hope* (Vol. 1) (N. Plaice, S. Plaice & P. Knight, Trans.). MIT Press.

Borges, J. L. (1941). *'The library of babel', in Ficciones*. Editorial Sur.

Brockington, D. (2009). *Celebrity and the environment: Fame, wealth and power in conservation*. Zed Books.

Bryman, A. (2004). *The Disneyization of society*. Sage.

Caillois, R. (1961). *Man, play and games* (M. Barash, Trans.). University of Illinois Press.

Carroll, L. (1865). *Alice's adventures in wonderland*. Macmillan.

Cassirer, E. (1944). *An essay on man: An introduction to a philosophy of human culture*. Yale University Press.

Chua, B. H. (2011). *Life is not complete without shopping: Consumption culture in Singapore*. Duke University Press.

Crawford, M. (1992). The world in a shopping mall. In M. Sorkin (Ed.), *Variations on a Theme Park: The new American City and the end of public space* (pp. 3–30). Hill and Wang.

Dillon, G. (Ed.). (2012). *Walking the clouds: An anthology of indigenous science fiction*. University of Arizona Press.

Eco, U. (1986). *Travels in hyperreality* (W. Weaver, Trans.). Harcourt Brace Jovanovich.

Eshun, K. (2003). Further considerations on Afrofuturism. *CR: The New Centennial Review, 3*(2), 287–302. https://doi.org/10.1353/ncr.2003.0021

Foucault, M. (1977). *Discipline and punish: The birth of the prison* (A. Sheridan, Trans.). Pantheon.

Foucault, M. (1980). In C. Gordon (Ed.), *Power/knowledge: Selected interviews and other writings, 1972–1977*. Harvester Press.

Foucault, M. (1991). Governmentality. In G. Burchell, C. Gordon, & P. Miller (Eds.), *The Foucault effect: Studies in governmentality* (pp. 87–104). University of Chicago Press.

Freud, S. (1919). The "uncanny". *Imago, 5*(5–6), 297–324.

Garland-Thomson, R. (2011). *Staring: How we look*. Oxford University Press.

Hannigan, J. (1998). *Fantasy City: Pleasure and profit in the postmodern Metropolis*. Routledge.

Haraway, D. J. (2008). *When species meet*. University of Minnesota Press.

Haraway, D. J. (2016). *Staying with the trouble: Making kin in the Chthulucene*. Duke University Press.

Harvey, D. (2000). *Spaces of hope*. Edinburgh University Press.

Hooks, B. (1992). *Black looks: Race and representation*. South End Press.

Husserl, E. (1970). *The crisis of European sciences and transcendental phenomenology* (D. Carr, Trans.). Northwestern University Press.

Kuhn, T. S. (1962). *The structure of scientific revolutions*. University of Chicago Press.

Lovecraft, H. P. (1928, February). The call of cthulhu. *Weird Tales*.

Massumi, B. (2002). *Parables for the virtual: Movement, affect, sensation*. Duke University Press.

Mbembe, A. (2003). Necropolitics. *Public Culture, 15*(1), 11–40. https://doi.org/10.1215/08992363-15-1-11

Miéville, C. (2009). *The city & the city*. Macmillan.

Pine, B. J., II, & Gilmore, J. H. (1999). *The experience economy*. Harvard Business School Press.

Sanderson, B. (2010). *The stormlight archive*. Tor.

Sontag, S. (1977). *On photography*. Farrar, Straus and Giroux.

Sorkin, M. (Ed.). (1992). *Variations on a theme park*. Hill and Wang.

Thrift, N. (2008). *Non-representational theory: Space, politics, affect*. Routledge.

Urry, J., & Larsen, J. (2011). *The tourist gaze 3.0*. Sage.

Vidler, A. (1992). *The architectural uncanny: Essays in the modern unhomely*. MIT Press.

Wasko, J. (2001). *Understanding Disney: The manufacture of fantasy*. Polity.

Whyte, K. (2018). Indigenous science (fiction) for the Anthropocene. *Environment and Planning, 1*(1–2), 224–242. https://doi.org/10.1177/2514848618777621

Wood, R. E. (2000). Caribbean cruise tourism. *Annals of Tourism Research, 27*(2), 345–370. https://doi.org/10.1016/S0160-7383(99)00073-0

Yusoff, K. (2018). *A billion black Anthropocenes or none*. University of Minnesota Press.

Žižek, S. (1991). *Looking Awry*. MIT Press.

Žižek, S. (1997). *The plague of fantasies*. Verso.

CHAPTER 4

Myth in the Landscape Bones: Landscape of Mind and Meaning

Abstract Mythical landscapes are treated as environments where mind and meaning take form through terrain, not as abstract symbols floating above place. Mountains, forests, deserts, islands, and caves recur across myth, philosophy, and popular culture as sites where cosmology is anchored, identity is tested, and ethical orientation is staged. Drawing on ideas of sacred geography and archetypal structures, these landscapes appear as enduring cultural forms through which societies organise awe, fear, transcendence, and transformation and insist that mythic terrains are also lived and sensed. Here, landscape is approached as a processual field made through movement, memory, and practice, where meaning accumulates through ritual, labour, and relational attention rather than through static representation. The analysis maps how particular terrains generate distinct mythic logics. Mountains function as vertical thought, linking ascent to revelation, discipline, and mastery, while also exposing the gendered and political assumptions embedded in elevation narratives. Forests operate as entangled worlds of concealment, haunting, and reciprocity, where spirits, animal agencies, and ecological intelligence unsettle human sovereignty and linear time. Deserts produce knowledge through exposure, scarcity, and horizon, but also carry colonial histories of "emptiness" that enable sacrifice zones, extraction, and nuclear testing. Islands intensify myth as experiment: bounded worlds that promise renewal and paradise while enabling control, surveillance, and epistemic capture. Caves register descent as a form of knowing, where enclosure, darkness, and

J. Coghlan and H. Nolan, *Landscapes That Speak*,
https://doi.org/10.1007/978-3-032-21757-8_4

bodily vulnerability complicate philosophical traditions that equate truth with exit and light. Across these terrains, myth is shown as a living method for making worlds intelligible under ecological crisis and postcolonial pressure, while also revealing how sacred meanings are folded into modern spectacle, governance, and dispossession.

Keywords Myth · Cosmology · Sacredness · Archetype · Dwelling · Sovereignty · Descent

Mythical landscapes are approached here as terrains saturated with symbolic resonance, spiritual potency, and narrative force, while also understood as lived, embodied environments rather than abstract backdrops. Drawing on Mircea Eliade's formulation of *sacred geography* and Carl Jung's analysis of *archetypal symbolism*, natural formations such as forests, mountains, islands, and deserts recur across cultures as sites of moral orientation, spiritual revelation, and psychological transformation (Eliade, 1959; Jung, 1964). Landscape phenomenology further reframes these terrains as places encountered through movement, memory, and sensory engagement. As Edward Casey argues, places actively participate in the formation of meaning and identity, shaping how worlds are felt and remembered rather than merely represented (Casey, 1997). This phenomenological perspective unsettles readings of mythical landscapes as static relics or romanticised imaginaries. Instead, such landscapes operate simultaneously as symbolic frameworks and lived environments. Tim Ingold's concept of the *taskscape* highlights their processual nature, emphasising how landscapes emerge through ongoing practices and relations between human and nonhuman actors (Ingold, 2000).

Mythical landscapes thus articulate ethical imperatives and existential concerns not as fixed symbols but as relational forms continually re-enacted through dwelling, ritual, and narrative. In this sense, they encode narratives of cosmological origin, rites of passage, ecological interconnectedness, and moral awakening while remaining responsive to changing historical conditions (Eliade, 1959). Crucially, mythical landscapes do not merely reflect cultural meanings but actively generate and renew them, infusing contemporary experiences of ecological crisis, postcolonial identity, and existential precarity with symbolic depth and ethical force (Harvey, 2006). They function as mediating terrains between inherited traditions and present-day anxieties, becoming key sites for negotiating belonging, ecological responsibility, and spiritual connection in the Anthropocene (Chakrabarty, 2021). Read alongside Indigenous cosmologies that foreground relationality and reciprocity (Kimmerer, 2013) and *speculative ecological* thought that challenges human exceptionalism (Haraway, 2016), mythical landscapes emerge as resources for resisting reductive narratives of disenchantment and extraction, offering instead alternative imaginaries grounded in relational ethics, ecological humility, and more-than-human interconnectedness.

The Mountain in Myth and Modernity

Born from an egg on a mountain top
The punkiest monkey that ever popped
He knew every magic trick under the sun
To tease the Gods,
And everyone and have some fun (Monkey Magic, 1978)

From ancient myth to modern philosophy, the mountain has served as one of humanity's most enduring symbolic landscapes. Its height creates a natural orientation point, inviting people to imagine the world in vertical terms; above and below, ascent and descent, order and limit. For Mircea Eliade, the mountain embodies the *axis mundi*, the point where different planes of existence meet and where the sacred becomes a material feature of the world (Eliade, 1959). High places are therefore not simply elevated terrain, they anchor cosmology. Mountains locate societies within a meaningful universe by binding ideas of heaven, earth, and the underworld along a single vertical line. Giambattista Vico explains the mountain's symbolic force as emerging from the earliest operations of metaphor. For Vico, the first humans, unable to think abstractly, animated the landscape with emotion and intention, turning mountains into living agents of power and meaning (Vico, 1744). The mountain became a fable in itself; a material form interpreted through the same imaginative faculty that produced myth and divinity. Kant later transforms this poetic animation into cognition when the mountain exceeds the mind's capacity to represent it, reason restores order through the experience of the *sublime* (Kant, 1790). What was once a sacred alignment becomes a cognitive structure; the mountain's height becomes a measure of the mind's confrontation with its own limits.

Hans Vaihinger extends this into a broader idea of thought as *as-if* construction, arguing that humans rely on fictions to orient themselves in a world too complex to apprehend directly. Under this logic, the mountain becomes less a divine revelation than a conceptual stabiliser, a landscape that helps thought organise scale and relation. Jung internalises this vertical logic even further; the hierarchy once mapped onto the cosmos becomes the architecture of the psyche, with height signalling the relation between consciousness and the unconscious (Jung, 1964). What was external becomes internalised as a model for psychic integration. Modern critical theorists trace the social implications of this ascent imagery. Marcuse sees the reconciliation of inner conflict as mirroring the managed stability of industrial society (Marcuse, 1955). This is a *monotheism of consciousness*, which enshrines unity, detachment, and control as ideals (Hillman, 1975). Rose (1993) identifies similar vertical patterns in gendered narratives of transcendence, showing how elevation, ascent, and distance from the everyday are repeatedly coded as masculine modes of mastery, authority, and spiritual overcoming, while horizontality, immanence, and proximity are feminised and aligned with care, embodiment, and relationality. Within mythical landscapes, these vertical imaginaries structure not only spatial hierarchies but also moral and epistemological ones, shaping who is imagined as capable of transcendence, vision, and control over nature.

Jameson (1981) notes how cultural forms translate historical tensions into psychological resolutions through images of rising above material constraint. Hence, the mountain persists not simply as a physical form but as a structure of thought. Its cultural forms translate historical tensions into psychological resolutions through recurring images of rising above material constraint. Within this logic, elevation operates as an imagined solution to social and material contradiction, offering transcendence in place of structural change. The mountain therefore persists as more than a physical form but as a structure of thought, naturalising ideals of mastery, distance, and moral superiority while displacing historical conflict into spatial metaphor. Mountain landscapes therefore can organise the mind, society, and meaning. The mountain's vertical logic survives as the architecture through which modern consciousness continues to orient itself. Mountains perform narrative work that few other landscapes can achieve.

Mythic Elevation: The Mountains Birth and Rebellion

The mountain's oldest narrative role appears not in geography but in mythology, where peaks mark the places where worlds meet and where beings are altered by proximity to the sky. The story of Huāguǒ Shān, "Flower and Fruit Mountain," in *Journey to the West* (Wu, 1592 [trans. Yu 2012]) remains one of the most compelling articulations of the mountains narrative function. Here, the mountain is not simply a location but a condensed cosmology, a meeting point of stone, cloud, water, and breath where geological matter becomes spirit. It is from this mountain that Sūn Wùkōng emerges, cracking open from a stone egg charged with celestial and earthly essence. His mineral origin collapses the boundary between matter and consciousness, marking him from birth as a being shaped by landscape rather than lineage. His early leaps across valleys, his time training atop remote peaks, and his ascent to the celestial court all unfold through mountain logic: to rise is to become. Wùkōng's repeated transformations, from stone to life, disciple to rebel, servant to seeker, replay the mountain's narrative rhythm of differentiation, striving, collapse, and renewal. Each ascent invites rebellion; each fall demands reinvention.

Modern retellings preserve this mountain-as-cosmos logic. In *Black Myth: Wukong* (2024), Huāguǒ Shān is rebuilt as an interactive world where mythic elevation becomes playable. Cliffs, waterfalls, temples, and cloud-wreathed summits generate a narrative architecture in which ascent is not merely movement but revelation. The mountain becomes a world-making engine, situating Wùkōng's story in a landscape that remains the primary agent of transformation. This logic echoes throughout culture. In *Moana* (2016), Te Fiti's volcanic heart becomes the centre of creation and return. In *Kung Fu Panda* (2008), the Jade Palace atop a precipice signifies enlightenment available only through discipline and ascent. In *Avatar: The Last Airbender* (2005–08), the Air Temples perched in high places generate entire philosophies of detachment, perspective, and freedom. Across these examples, the mountain's narrative role endures as a birthplace of power, the site of rebellion, and the arena of becoming. Other narratives deepen this pattern. In *The Lion King* (1994), Pride Rock functions as a secular mountain of revelation: births, blessings, and exiles play out on its summit, where height situates characters within cycles of life and governance.

In *Hercules* (1997), Mount Olympus stands as the aspirational summit of divine belonging; ascent marks the hero's transition from mortal to god, and descent marks his willingness to choose humanity over celestial privilege. In the indie videogame *Journey* (2012), a distant mountain peak is the story's constant visual anchor, its glowing summit pulling the player across deserts, ruins, and snowfields. Every movement is oriented toward the mountain, which becomes a quiet yet commanding narrative magnet. Even in literary traditions mountains structure identity. In Mary Shelley's *Frankenstein* (1818), Victor's encounters in the Alps provide clarity and rupture, the scale of the mountains dissolves his sense of personal grievance and reorients him toward larger forces. Jack Kerouac's *Dharma Bums* (1958) frames mountain ascent as spiritual aspiration, where the peak becomes a place of temporary illumination and fleeting peace. Chimamanda Ngozi Adichie's *Purple Hibiscus* (2003) uses the Nsukka hills as a site of awakening, where elevation grants distance from domestic oppression and creates the conditions for self-discovery.

Mountains as Cinematic Machinery: Spectacle, Threshold, and Narrative Scale

Cinema transforms mountains into dynamic narrative instruments, using their scale to shape pacing, atmosphere, and revelation. Because the mountain alters how the camera moves requiring aerial sweeps, long shots, steep angles, and vertical framing it actively determines the story's visual language. *Baraka* (1992) non-narrative, non-verbal documentary film. It contains no dialogue and no voice-over, instead, it is structured as a visual and musical meditation on human cultures, rituals, landscapes, and the more-than-human world. Filmed across over 20 countries and shot in 70 mm film using time-lapse, slow motion, and aerial cinematography it captures sacred rituals, natural landscapes, and scenes of modern life through striking, wordless imagery. Using sweeping panoramic shots, the film juxtaposes human activity with the planet's vastness to explore humanity's relationship to the Earth. In particular, its sweeping shots of the Himalayas present the mountain as *sublime* presence, but the film's framing inevitably reshapes immensity into something visually consumable. What once felt ungraspable becomes aesthetic, a stillness suspended in cinematic time.

Peter Jackson's *The Lord of the Rings* trilogy (2001–03) deepens this logic. Mountains function as narrative thresholds. Caradhras refuses passage through avalanches that feel almost intentional, redirecting the Fellowship into the darkness of Moria. Inside the mountain, vast halls and bottomless chasms reveal the ruins of a civilisation undone by its own extraction: "The dwarves delved too greedily and too deep." Here, the mountain becomes a narrative judge, releasing shadow and flame as consequence. Mount Doom is the trilogy's narrative anchor, a peak whose every appearance condenses tension and signals the story's gravitational centre. As the physical destination of the quest and the symbolic heart of Middle-earth's corruption, its volcanic form structures the unfolding drama, generating dread even when it is distant or unseen. Jackson's pairing of close, intimate shots of Frodo with sweeping, smoke-wreathed vistas of the mountain compresses vast scale into emotional proximity, allowing the landscape's hostility and enormity to register directly on Frodo's body and mind. In this way, the external ascent toward Mount Doom becomes inseparable from the internal unravelling it provokes, binding geography and psychology into a single escalating narrative pressure.

Videogames also mobilise mountains not simply as scenery but as structuring devices that organise how players perceive, move through, and understand narrative space. In *The Legend of Zelda: Breath of the Wild* (2017), peaks are more than navigational challenges; they shape the game's entire grammar of exploration. Every climb produces a new sightline, allowing the player to grasp the world's geography in sweeping, revelation-like moments. The cartographic clarity of a high vantage point becomes a narrative event: reaching a summit rewrites the player's understanding of the world's scale, dangers, and future paths. Mountains thus function as story nodes, places where the plot dilates or changes direction because the player sees differently. Altitude becomes knowledge. Visibility becomes agency. In *Celeste* (2018), the mountain is internal rather than external. Its shifting terrain of disappearing platforms, mirrored chambers, and unpredictable winds externalises psychological turbulence. Climbing the mountain becomes a process of encountering fear, grief, anxiety, and resilience in embodied form; each level exposes an emotional fault line and demands a new form of adaptation. The mountain grows increasingly surreal as the narrative intensifies, suggesting that elevation amplifies the self rather than liberating it from burden. Where other games treat mountains as goals, *Celeste* transforms ascent into emotional cartography: the peak is not a destination but a confrontation with the internal landscapes that shape identity.

Shadow of the Colossus (2005) pushes the mountain logic further by collapsing creature and terrain into one form. Many colossi are essentially living mountains, vast bodies of stone and fur that mimic cliffs, ridges, and plateaus. Climbing them resembles ascending a sacred peak: uncertain footing, sudden exposure, shifts in gravity, and the ever-present risk of falling. The game renders ascent as both physical exertion and ethical dilemma. The mountain-beings are neither enemies nor simple obstacles, they feel ancient, mournful, and tethered to the landscape. Scaling them becomes a mythic trespass, an act that disturbs the balance of a world built on quiet, monumental stillness. Here, mountains are not passive structures but sentient sites, embodiments of a world's memory and order. Other games extend this mountain logic in diverse ways. In *Skyrim* (2011), the Throat of the World serves as narrative anchor and spiritual summit, a place where dragons speak the language of creation. In *Death Stranding* (2019), mountain ranges recalibrate pacing entirely: snowy ascents slow movement to a crawl, force strategic planning, and create narrative tension through exposure and isolation. *Journey* (2012) builds their entire storytelling arc around a distant mountain peak, using its slow, looming presence as a visual and emotional compass. Across these forms, the mountain becomes a narrative architect, shaping how stories unfold through verticality, scale, risk, and revelation. Videogames, more than any other medium, arguably allow players to *feel* the mountain's narrative work through embodied control; straining to climb, slipping, reaching plateaus and seeing the world expand beneath them. Mountain landscapes in video games thus become worlds in themselves, terrains where story and environment merge, and where ascent is both an action and an unfolding meaning.

Mountains as Existential and Ecological Encounters

Not all mountain narratives seek spectacle or mastery. Many turn away from triumphalist logic and instead frame the mountain as a place of confrontation, stillness, and exposure as landscapes where human identity undergoes reduction rather than elevation. In these works, the mountain strips away narrative excess. It becomes a terrain where meaning thins to elemental essentials, where characters confront not what the mountain *is*, but what they *are* in its presence. Werner Herzog's *The Dark Glow of the Mountains* (1984) exemplifies this approach. Rather than celebrating Reinhold Messner and Hans Kammerlander's attempt to climb Gasherbrum I and II, Herzog focuses on the interior implications of high-altitude climbing. He refuses summit footage, the usual climax of mountaineering films. The camera lingers instead on windless plateaus, long silences, and the tenuous rhythm of breath at altitude. Messner's reflections reveal the mountain not as a challenge to overcome, but as a presence that reorganises perception. Gasherbrum is treated as an indifferent world unto itself, a landscape whose vastness exposes the contingency of human ambition. Its slopes serve as mirrors in which fear, desire, ego, and humility reflect without distortion.

Touching the Void (2003) deepens this existential register through narrative reconstruction. Based on Joe Simpson's memoir, the film recounts a climb in the Peruvian Andes that becomes a descent into psychological and physical extremity. When Simpson falls into a crevasse and is left for dead, the mountain becomes a crucible of hallucination, despair, and the will to survive. The landscape is stripped of romance; whiteout conditions erase horizon and depth, producing a kind of sensory annihilation. The mountain's indifference becomes palpable. Survival arises not from conquering the peak but from navigating a world where orientation, identity, and even the notion of time collapse. The mountain becomes the medium through which meaning dissolves and is painfully remade. In *Free Solo* (2018), the camera's intimacy with El Capitan's granite face transforms the mountain into a psychological surface. Alex Honnold's ascent without ropes externalises internal conflict; the sheer wall reveals his perfectionism, emotional restraint, and deep-seated fear of inadequacy. Close shots of fingertips on microscopic holds reveal the body's negotiation with stone. El Capitan is depicted with dignity and indifference; it neither aids nor resists. It simply exists, and Honnold must find a place within *its* geometry. Here, the mountain becomes an ecological encounter, a relationship that resists instrumentalisation, demanding attention rather than mastery.

Other works further develop this existential and ecological terrain. *North Face* (2008), based on the 1936 Eiger climb, transforms the North Wall into a stage of political pressure and bodily devastation, revealing how the mountain absorbs and reflects human ideologies. *Everest* (2015) revisits the 1996 disaster by foregrounding weather, altitude sickness, and logistical failure, rendering the mountain as a force that reshapes human intention without malice. Even earlier films like *The Eiger Sanction* (1975) contain moments where the mountain's indifference pushes against narrative conventions, hinting at a landscape more complex than the spy-thriller plot allows. More contemplative works such as *Mountain* (2017) and its companion piece *Sherpa* (2015) reclaim the mountain as a space of human–nonhuman entanglement. Through sweeping cinematography and reflective narration, *Mountain* emphasises how peaks inspire awe while revealing the tension between ancient landscapes and modern appetites for risk and conquest. *Sherpa* reframes Everest through the eyes of the Sherpa community, showing how mountains are not blank canvases but embedded within social, cultural, and ecological relationships that Western narratives often erase.

Documentaries like *Meru* (2015) and *The Alpinist* (2021) deepen this relational view. The hostile beauty of Meru Peak becomes inseparable from the interpersonal dynamics and psychological scars of its climbers. Marc-André Leclerc, the subject of *The Alpinist*, treats mountains with an almost animist respect viewing them not as stages for achievement but as worlds whose rhythms he must match. His climbs become acts of listening. The mountain dictates mood, weather, and tempo. The narrative is shaped by the landscape's shifting permissions. These existential mountain narratives collectively challenge the blockbuster model where mountains are framed as resources, obstacles, or systems of calculable danger. Instead, they foreground the mountain as agentive presence, a terrain that shapes thought, perception, and possibility simply by existing. Peaks are not conquered, they are encountered. They are not antagonists, they are conditions. Their indifference opens a space in which narratives can register vulnerability, humility, clarity, or dissolution.

Forests as Mythic and Narrative Landscapes

> In ancient times, the land lay covered in forests, where, from ages long past, dwelt the spirits of the gods ... Those that remained were guarded by gigantic beasts ... For those were the days of gods and of demons. (Narrator, *Princess Mononoke* 1997)

Forests absorb and create meaning differently from mountains. Where the mountain imposes orientation through verticality, height, and ascent, the forest dissolves orientation through horizontal entanglement. Its logic is not vertical but diffuse, a shifting weave of thresholds, shadows, and recursive paths that generate imagination, danger, and change. From the earliest sacred groves to contemporary ecological discourse, the forest represents both refuge and entrapment, a space where thought loses its coordinate system and meaning becomes relational. Across religious and mythological traditions, sacred forests form part of a broader cosmological grammar, where trees and groves act as mediating structures that connect human, divine, and nonhuman worlds. In Greece, the oak grove of Dodona spoke through rustling leaves and trembling air, translating divine presence into atmospheric sensation. In Norse cosmology, Yggdrasil held the nine worlds together as a living scaffold of existence, its roots and branches threading underworld, earth, and sky. Under the Bodhi Tree, the Buddha attained enlightenment, revealing the tree not as a ladder to the heavens but as a site of immanent illumination. In each tradition, forested landscapes are not merely settings but media through which divine, earthly, and human forces circulate in continuity. Modernity often internalises these spatial relations, translating external cosmological networks into metaphors of interiority. What once functioned as a living, relational system of immersion and return became symbol, allegory, or psychological depth. Yet forests, as well as mountains, can also be understood as material and semiotic fields in their own right: environments that generate meaning alongside and beyond human perception, sustaining cosmological resonance even as their symbolic registers shift.

Hayao Miyazaki's cinema returns again and again to the forest as a place of ancient intelligence, ecological reciprocity, and mythic presence. Miyazaki is a Japanese filmmaker, animator, and co-founder of Studio Ghibli, widely regarded as one of the most influential figures in world cinema and animation. His animated feature films combine meticulous hand-drawn artistry with sustained philosophical, ecological, and ethical inquiry, consistently centring on landscape, movement, and human–more-than-human relations. In Miyazaki's work, landscapes function as living, sentient environments rather than passive backdrops, shaping narrative possibility and moral orientation through forests, rivers, winds, ruins, and mountains animated by Shinto cosmologies in which kami inhabit both natural and built worlds. Miyazaki's cinema is marked by a persistent critique of war, industrialisation, and extractive modernity, articulated through narratives that resist mastery and technological triumphalism. Ethical insight emerges instead through embodied labour, care, and everyday practice, while non-teleological storytelling privileges wandering, process, and becoming over conquest or resolution. Read through phenomenological and environmental frameworks, Miyazaki's films operate as cinematic landscapes of ethical encounter, where attention, humility, and relational engagement replace domination as the conditions for inhabiting complex ecological worlds. His work captures a simple but profound cosmological insight, in this case, where the forest is where worlds meet.

His work captures a simple but profound cosmological insight: the forest is the space where worlds meet. Miyazaki's *Princess Mononoke* (1997) follows Ashitaka, a young prince cursed while defending his village from a demonised boar god. Seeking a cure, Ashitaka becomes caught in a violent conflict between Iron Town, led by Lady Eboshi, and the surrounding forest inhabited by animal gods and spirits, including San, a human girl raised by wolves. As human industrial expansion threatens the forest's survival, the struggle culminates in the beheading of the Forest Spirit, unleashing ecological chaos. The film rejects simple moral binaries, instead exploring coexistence, interdependence, and the ethical limits of human progress. Here, the forest operates as a liminal threshold between human settlements, animal deities, and ancient ecological forces, dissolving clear boundaries between civilisation and nature, spirit and matter. Rather than functioning as a sanctuary removed from conflict, the forest becomes a contested, relational space in which ethical responsibility emerges through encounter, negotiation, and restraint.

Richard Powers' novels especially *The Overstory* (2018) and *The Echo Maker* (2006) reimagines the forest as a form of narrative intelligence: a distributed system of perception, memory, and communication that challenges human-centred understandings of meaning. Powers' novels do not treat trees as metaphors for human consciousness, rather they present forests as agents that create relation, structure narrative, and reshape the temporal scale of the story itself. Eduardo Kohn names this type of entanglement an *ecology of selves*, a communicative environment where representation extends beyond human thought. Narrative works frequently use forests to explore forms of relational life, and Ursula K. Le Guin's *The Word for World Is Forest* (1972) remains one of the most influential examples in modern speculative fiction. The novella imagines the planet Athshe, whose dense green canopy shapes the culture, cognition, and cosmology of its Indigenous inhabitants, the Athsheans. In this world, the forest is not a backdrop for human action but a living, relational environment that structures every dimension of existence. It is the medium through which thought occurs, the organising principle of society, and the primary source of meaning. Le Guin uses this forested world to challenge Western assumptions about agency, perception, and power. For the Athsheans, the forest is neither property nor resource, instead it is home, teacher, archive, and dreamspace; a landscape that sustains them materially and spiritually. The Athsheans' worldview emerges from the rhythms of ecological reciprocity that bind them to their forest environment. Within this relational cosmology, dreaming and waking form a continuous spectrum of awareness. Athshean knowledge develops not through domination or extraction but through cycles of growth, rest, repetition, and return, a cognitive ecology shaped by the forest's own tempo. Le Guin presents a society whose understanding of reality arises from participating in the forest's rhythms, offering an alternative model of intelligence grounded in connection rather than control. Similarly, *Princess Mononoke* frames the forest as a breathing field of wound and renewal. Every act of harm alters both the physical and spiritual order. The Great Forest Spirit embodies nature's capacity for both destruction and fertility, showing that no action escapes consequence.

Philosophically, Næss sees this interrelation as the basis of deep ecology (Næss, 1973). For him, the self expands by participating in the whole field of life rather than standing apart from it. A person's identity grows when they recognise their dependence on a larger web of relations, and ethical responsibility emerges not from ownership or control but from identification by understanding oneself as part of that continuity (Næss, 1989). Donna Haraway reframes this as *sympoiesis*, the co-making of life through material reciprocity. Her concept of *tentacular thinking* mirrors forest logic: being emerges through partial connections rather than isolated ascent. Forest ecologies model this pattern. Forest ecologies operate through forms of mediated entanglement rather than direct contact. Mycorrhizal networks link trees and plants via fungal systems that transfer nutrients, signals, and stress responses without roots or canopies ever touching, producing a distributed, relational ecology often described as the *wood wide web* (Simard, 1997; Simard et al., 2013). *The Word for World Is Forest* enacts this compositional logic through shifts between dreaming and waking, emphasising knowledge as rhythm rather than revelation. The narrative mimics the metabolism of the forest: ideas germinate, spread, and return in shifting patterns that accumulate meaning through relation.

Contemporary cinema and digital worlds extend this logic. In *Annihilation* (2018), species exchange genetic material through a shimmering field that dissolves boundaries of identity meaning the landscape becomes a site of recombination. Video games such as *The Witcher 3* and *The Legend of Zelda: Tears of the Kingdom* translate forest intelligence into environmental design. Mechanics like Ultrahand and Fuse encourage players to treat the environment as collaborator rather than resource, progress emerges through attunement to latent relations among wood, stone, wind, and water. The forest's tentacular logic becomes playable cognition. Digital simulation reveals the fragility of the organic systems it attempts to copy. Songs such as Radiohead's "Fake Plastic Trees" (1995) and Joni Mitchell's "Big Yellow Taxi" (1970) register this loss, lamenting the replacement of living, interconnected worlds with artificial and manageable substitutes.

Forests in Folklore and Fairy Tale: Grimm, Perrault, and Angela Carter

Across European folklore and fairy tale traditions, the forest is a consistent narrative landscape. It appears as a zone of danger, enchantment, testing, and transformation. It is a mythic terrain where ordinary social rules dissolve and deeper truths emerge. From the oral tales collected by the Brothers Grimm to Perrault's literary fairy tales and Angela Carter's feminist rewritings, the forest remains the quintessential space of narrative metamorphosis. Cinema and television have amplified this symbolic power, using framing, lighting, and sound to restore the forest's mythic resonance. In the Grimm tradition, the forest is a threshold space, marking the boundary between the village as a site of social order and the wilderness, where human identity is tested. It is the place where children are abandoned (*Hansel and Gretel*), where paths become misleading (*Little Red Riding Hood*), and where metamorphosis is possible (*The Six Swans*). The forest in these tales is not simply a backdrop but a narrative machine, which produces fear, isolates characters, and prompts ingenuity or moral growth. Early cinematic adaptations emphasised the forest's darkness and instability. Classic animations like Disney's *Snow White and the Seven Dwarfs* (1937) deploy expressive, exaggerated imagery of trees with twisting faces, grasping branches, and enveloping shadows to externalise the heroine's fear. The cinematic frame becomes a psychological mirror. Unlike earlier discussion, here the forest is not real, it is a manifestation of interior states. This is a key feature of mythical forest landscapes: they can collapse the distinction between psychological and physical terrain.

Charles Perrault's tales from the late seventeenth century, such as *Little Red Riding Hood*, use the forest not only as a dangerous place but also as a site of social instruction. Perrault frames the forest as a moral warning, particularly for young women. Here, the forest is less wild landscape than allegorical stage, a space where the reader learns the consequences of straying from prescribed paths. Cinematic retellings often amplify the sensuality and ambiguity latent in Perrault's stories. Films like *The Company of Wolves* (1984), inspired by both Perrault and Angela Carter, rework *Little Red Riding Hood* into Rosaleen, a young girl living in a forested village who hears a series of cautionary stories from her grandmother about wolves, men, and the dangers that emerge at the threshold between childhood and adulthood. These tales, embedded as dreamlike vignettes, blur the boundary between folklore, fantasy, and psychological warning. Central to them is the idea that wolves are not only animals but men who reveal their true nature when desire, hunger, or social rules are loosened. As Rosaleen ventures into the forest, she encounters a charming huntsman who later reveals himself to be a werewolf. Rather than ending in simple victimhood or rescue, the film culminates in Rosaleen's transformation, as she joins the wolves instead of fearing them. The forest becomes a space of sexual awakening, danger, and self-recognition, where rigid moral binaries collapse. The film reframes the forest as a landscape in which identity, desire, and power are negotiated rather than resolved while also using deep reds, stylised fog, and symbolic framing to reveal how forest spaces encode cultural anxieties about desire, agency, and transgression. The camera lingers on the moment the path vanishes into shadow, an image that captures the forest's narrative purpose as the site where innocence is questioned, and boundaries dissolve. Angela Carter revitalises the forest in her *Bloody Chamber* tales (1979), transforming it into a terrain of empowerment rather than punishment. In her hands, the forest becomes a space where women renegotiate identity and reclaim agency. Instead of being devoured or rescued, Carter's protagonists confront monsters, rewrite scripts, or embrace their own wildness. This shift is deeply tied to landscape. Carter's forests throb with sensuality and possibility. They are not warnings but invitations. The forest no longer externalises fear instead it externalises desire, appetite, and autonomy. Carter's adaptations, including Neil Jordan's *The Company of Wolves* (1984), use lush cinematography and dreamlike sequences to explore this revised symbolic geography.

Forests in Horror: The Witch, The Blair Witch Project, and Twin Peaks

In horror cinema and television, the forest becomes a mythic landscape of fear, disorientation, and the collapse of rational order. Unlike the enchanted forests of folklore, which promise transformation or moral learning, horror forests resist comprehension altogether. They are territories where human perception breaks down, where the familiar dissolves into the uncanny, and where the landscape itself appears sentient. Through cinematic framing, sound design, and narrative architecture, horror transforms the forest into a character, an agent of terror rather than a passive setting. Robert Eggers' *The Witch* (2015) uses the seventeenth-century New England forest to embody Puritan fears of wilderness and spiritual corruption. From the family's first expulsion from their plantation, the forest is framed as a dark perimeter: a boundary marking the limits of both Christian worldview and patriarchal control. Close framing and muted colour palettes render the forest oppressive and unknowable, its tangled branches forming visual barriers that mirror the family's growing paranoia. The camera rarely reveals the forest's depths, instead it hints, obscures, and distorts. This refusal to fully show the woods maintains their mythic function: the unknowable can't be mapped, only feared. In Eggers' film, the forest is not merely outside civilisation, it is the place civilisation defines itself against.

The Blair Witch Project (1999) demonstrates the forest functioning as an active narrative force rather than a passive setting. Shot in a found-footage style, the film embeds its horror within the mechanics of representation, using the camera's limitations to reveal how the forest destabilises perception, orientation, and meaning. The forest becomes a labyrinth of repetition: the same stream crossed multiple times; the same log encountered after hours of walking. Here, the horror arises from spatial betrayal because the forest refuses to obey rational mapping, and the camera becomes a witness to the dissolution of coherent space. The forest is framed almost entirely through shaky, handheld shots that deny viewers a stable perspective. This technique is crucial as it makes the forest feel alive; twigs break without a visible agent, stone piles appear overnight, figures are glimpsed only in peripheral vision. The absence of a visible monster reinforces the idea that the forest itself is the antagonist. The Blair Witch becomes the forest's rumour, its mythic consciousness externalised through sound, movement, and misdirection.

The landscape becomes the narrative engine, producing disorientation and terror through form alone.

In David Lynch's *Twin Peaks* (1990–91; 2017), the forest of the Pacific Northwest is a mythic landscape of haunting and revelation. Unlike conventional horror films, which localise threat in a monster or witch, *Twin Peaks* locates the supernatural within the forest's very atmosphere. Slow tracking shots through mist-filled trees, repeated motifs of wind moving through branches, and the constant hooting of owls establish the woods as a portal, a site where dimensions overlap and where the boundaries between inner and outer worlds blur. The forest houses the Black Lodge, a space of psychic and metaphysical significance. It is not simply a haunted location but a symbolic architecture of trauma, desire, and violence. Characters enter the forest and emerge changed or destroyed, Lynch's use of framing including lingering on dark openings between trees and using slow zooms to mimic hypnotic pull turns the forest into a visual metaphor for the unconscious. It is a mythic psyche made geographical.

The Forest as Narrative Landscape: Weirwoods, Memory, and Mythic Frontiers in *Game of Thrones*

In *Game of Thrones* (2011–19), the forest emerges as one of the most potent narrative landscapes, shaping the series' cosmology and structuring its deepest tensions. While the political centres of Westeros with its castles, courts, and battlefields anchor the drama of human ambition, the forests exist outside these systems of governance and chronology. They are places where time stretches, where older powers linger, and where human history brushes against forces that pre-date and exceed it. The forest functions not as a scenic background but as a mythic agent whose presence directs the movement of the story and transforms the characters drawn into its depths. At the heart of this forested world stand the weirwood trees. Their white trunks and blood-red leaves form a visual and symbolic contrast to the harsh northern landscape, signalling a realm where ordinary perception falters and the past remains materially alive. Each carved face gazing from a heart tree embodies ancestral presence; the weirwood is both shrine and witness. Through Bran Stark's visions, the series makes clear that memory in this world is not abstract but rooted. It flows through wood, water, and wind. The forest becomes an archive that cannot be rewritten, a narrative record carried in living matter rather than in texts. In this sense, the weirwoods function as a landscape, a point where worlds intersect and where the continuity of time is not conceptual but embodied. The forest preserves histories that political institutions struggle to contain, sustaining mythic truth in a terrain older than any dynasty.

Beyond the Wall, the forest shifts in register and becomes a mythic frontier. The Haunted Forest marks the place where the distinction between life and death grows uncertain, where the familiar landscape of Westeros gives way to something vaster and more primordial. Giants, direwolves, White Walkers, and wights (the supernatural or undead beings in folklore and myth, most commonly associated with Germanic and Norse traditions) appear not merely as supernatural creatures but as beings shaped by cosmological forces that human characters barely comprehend. The forest behaves like the enchanted woods of folklore, where travellers undergo trials that remake their sense of self, but it adds a starkly contemporary dimension: its dangers evoke climate catastrophe, extinction, and cyclical apocalypse. The Night King transforms the forest into an ice-bound stage on which the world's oldest conflict unfolds, revealing that the true stakes of the narrative lie not in political manoeuvring but in the cosmic rhythms of survival and annihilation. Across these varied environments, the forest shapes *Game of Thrones* at the level of narrative logic: It holds the oldest stories, predating the kingdoms and their wars; it reveals truths, often reluctantly, to characters who must learn to see beyond the limits of their social world; and it tests those who enter its depths, whether Jon Snow seeking leadership, Bran Stark seeking vision, or the Night's Watch seeking purpose. The forest articulates thresholds between the known and the ungraspable, drawing the narrative repeatedly to its edges to confront what human authority cannot contain. The forests of *Game of Thrones* act as a mythic counterpoint to the drama of power. They remind viewers that political ambition is fragile against the deeper orders of land, memory, and myth. Whether as sanctuary or menace, archive or frontier, the forest endures as the terrain where the world's most ancient truths reside. It is the place where power is stripped of illusion, where continuity asserts itself against human schemes, and where the story's most profound revelations are rooted in the living landscape itself.

Animal Umwelten and Forest Camouflage

The forest contains not a single reality but multiple overlapping perceptual worlds. Each organism inhabits a distinct field of awareness defined by its sensory reach and mode of response. Jakob von Uexküll first introduced this conception in his 1909 essay *Umwelt- und Innenwelt der Tiere* (*Environment and Inner World of Animals*), where he distinguished the *Umwelt* (the organism's subjective perceptual world) from the *Umgebung* (the objective surroundings) (Uexküll, 1909, 1–62). Every creature, he proposed, lives within a closed circuit of perception and action, constructing a world meaningful only within the range of its sensory organs and effectors. This subjective environment forms what he called a *sense-island*, a limited yet coherent field where each sign of perception connects directly to a sign of effect (Winthrop-Young, 2010, 239).

Camouflage demonstrates how these systems evolve through reciprocal influence. Texture shapes vision, and vision refines texture. The tree's bark develops fissures that absorb light and the moth's wing mirrors those patterns until predator and prey together define the limits of recognition. Evolution becomes a form of semiotic training where every adaptation both reads and rewrites the surrounding world. Form and perception co-construct the conditions of survival, turning the forest into a living archive of correspondence. Cinematic and digital representations reinterpret this semiotic exchange through human technologies of sight. Forests are often depicted as places where one can't trust their own eyes. It is an unforgivable break of the human covenant with reality, that seeing is believing. Cinema plays with light and shadow in forest animation in particular when it recast shadows as living writhing monsters. TV shows like *Lost* (2004–2010) and the *Jurassic Park* franchise (1993–2022) make use of the unseen with arial shots of trees shaking, or the sounds of something large enough to fell trees as it moves just out of view. Dolly's zoom down forest paths, combines a zoom-in on the lens with a simultaneous dolly-out (or the reverse), keeping the subject roughly the same size while distorting spatial depth. The cinematography of Frodo sensing the Ringwraiths in The *Fellowship of the Ring* (2001) makes the path appear to stretch or expand while the background warps in relation to the foreground. The result exaggerates perspective and creates an unsettling sense of spatial dislocation, often paired with wind or movement to heighten tension.

In *Predator* (1987), the alien hunter's invisibility manipulates the forest's perceptual logic rather than replacing it. The creature's cloaking device bends and redirects photons so that the viewer sees the actual trees and foliage behind the Predator as if no body interrupts the scene. The distortion produces a rippling heat-haze shimmer, blending the creature into the living texture of the forest. This raises a subtle philosophical and perceptual problem: if the eye registers photons that have been curved around an object instead of reflected from it, is the observer seeing the forest itself or a technologically mediated version of it? The Predator becomes a disturbance in the landscape rather than an absence but as presence defined by how it alters the viewer's vision of the environment. By contrast, the invisible car in *Die Another Day* (2002) is a completely different technological fiction. As Q explains, tiny cameras on all sides of the Aston Martin capture the surrounding environment and project those images onto a light-emitting surface, effectively creating a moving screen. In this case, the viewer is not seeing the world *through* the car via bent light, but rather a reproduction of the landscape displayed across its exterior. The car's invisibility works by simulation, not distortion: a curated image replaces the real world. Thus, the two forms of invisibility engage the landscape differently. Bond's car offers an image of the forest, a projected facsimile. The Predator offers a deformation of the forest, a bending of its light. Both appear invisible, but one erases itself with representation, while the other reveals itself only through the shimmering disruptions it causes in the world it is trying to hide within.

In nature, camouflage is an evolved dialectic. It is evidence of the predator–prey relationship as dictated by natural selection. The effectiveness of visual camouflage often depends on who is looking. Many ecologists have questioned how a red fox or a bright orange tiger can act as camouflage, at least until they realise the relative uniqueness of trichromatic vision in mammals. Once people realise many of the animals seeing foxes and tigers cannot differentiate between orange and green, the camouflage becomes obvious. The revelation often leads to the question: if humans can see more colours than other mammals, what does the world look like to those that can see in the ultraviolet spectrum. The mantis shrimp (order *Stomatopoda*) possesses one of the most complex visual systems known. Humans have three types of photoreceptor cells (cones) for colour detection (red, green, and blue) allowing trichromatic vision. The mantis shrimp has up to sixteen distinct photoreceptor types, including multiple ultraviolet and polarisation-sensitive receptors.

The cryptic impact of a limited perceptual field appears in *The Hunger Games: Catching Fire* (2013), where the arena's holographic dome projects an illusory sky and horizon onto its surface, creating the impression of an open landscape. The invisible barrier is not achieved by bending light, nor by projecting images onto an object, but by masking the physical structure of the dome with a seamless digital illusion. When Katniss shoots an arrow at the sky, the point of impact briefly reveals the projection system: the sky flickers, cracks appear in the simulation, and the invisible dome becomes perceptible through its failure. This kind of invisibility operates not by removing an object from perception, but by substituting the landscape itself with an image, replacing the real atmospheric space with a technologically maintained fiction. What the viewer perceives is not the forest canopy or open air, but a digitally imposed environment that conceals the arena's true form. Here, invisibility is produced not by distorting what is present, as in *Predator*, nor by projecting images onto an object, as in Bond's Aston Martin, but by overlaying false landscape onto real infrastructure. The moment the illusion falters, the artificiality of the world is exposed, revealing how invisibility can function as an architectural sleight of hand rather than a property of objects themselves.

Earlier films express the same displacement through different tones. *Bambi* (Disney, 1942) presents perception as innocence, its limited awareness organised for human recognition. *Bambi*, attempts to depict the deer's perception produces a human version of animal awareness. The film confines experience to sound, movement, and light, but these signals follow the structure of human expectation. Each sign of fear, safety, or attachment operates within a language of human emotion. The apparent immediacy of animal perception becomes a translation into recognisable form. Bambi, like many depictions of landscape on film is more akin to Bond's Aston Martin than the Predator. It is a representation of nature reproduced for human consumption. It is important that we don't allow it to camouflage the true nature behind it. Thomas Nagel's "What Is It Like to Be a Bat?" (1974) clarifies the structural limit that governs this translation. Consciousness depends on the organisation of its own senses, human vision cannot enter another sensory world. Every image of nonhuman perception repeats human cognition under another name. *Bambi* materialises this condition in cinematic form. It constructs the illusion of a deer's world while maintaining the distance that separates observer and observed; unrealities camouflaged as nature. Uexküll's *Umwelt* and

Nagel's analysis describe the same boundary through different vocabularies. Every being generates meaning within its perceptual range, and all attempts to extend that range return to the conditions of human sense. The forest holds these separate orders in simultaneous presence, yet representation cannot unify them. Cinematic and digital images of the animal and the forest appear as reality but remain shaped by human faculties. They are forms of camouflage in which mediation hides behind the appearance of immediacy, preserving the human frame while projecting it onto the living world.

Desert as Revelation: Sacred Geography and Its Discontents

I've been through the desert with a horse with no name
It was good to be out of the rain
In the desert you can't remember your name
Because there ain't no one for to give you no pain- A Horse with no Name
(Bunnell, 1971)

Desert landscapes occupy a fundamentally different position to forests. Rather than anchoring order, the desert lies at the margins of the created world, where coherence thins and boundaries become porous (Eliade, 1959). What *sacred geography* contains, the desert exposes. Its vastness resists enclosure, hierarchy, and orientation, confronting the subject with openness rather than centre. This exposure is not simply a condition of lack or deprivation but a potential site of revelation, where the stripping away of structure enables moments of clarity and encounter. This understanding resonates with phenomenological accounts of desert space. Yi-Fu Tuan describes deserts as landscapes of heightened awareness, where scarcity intensifies perception and draws the body into direct sensory engagement with light, heat, horizon, and wind (Tuan, 1990, 112). Edward Casey similarly argues that the absence of anchoring forms such as tree lines, built structures, or enclosing landmarks produces an intensified experience of place in which self, space, and narrative are pared back to their essentials. In this sense, the desert functions as a narrative aperture, an open terrain through which meaning, disorientation, and transformation become newly possible.

The sparseness of the desert is therefore not absence but generative minimalism. Its reduction of features creates a landscape where movement, perception, and survival become forms of narrative action. Tim Ingold's notion that landscapes are *tasks* rather than settings (Ingold, 1993, 152) is particularly resonant here. In desert narratives, simply existing in the environment becomes an activity, a negotiation with distance, exposure, and the shifting line where ground meets sky. Nothing in the desert is given and every orientation must be earned. This makes deserts powerful narrative landscapes. They do not conceal meaning rather they reveal it by subtraction. Simon Schama notes that desert iconography has long been associated with both revelation and ruin, functioning historically as a moral testing ground where individuals confront elemental truths stripped of cultural ornament (Schama, 1995). John Wylie adds that such terrains *undo* the viewer's sense of stability, replacing comfort with a sharpened awareness of vulnerability and scale (Wylie, 2007). In desert landscapes, narrative and perception converge because the desert reorganises the terms through which the world is apprehended. Seen through this broader lens, the desert becomes more than a place of spiritual austerity. It is a narrative mechanism, one that produces insight by reducing the world to horizon, heat, and endurance. In desert narratives, characters move across a landscape where structure dissolves into expanse, where form approaches formlessness, and where the possibility of transformation becomes visible precisely because so little else remains.

Deserts as Narrative Landscapes: Knowledge and Discipline

Deserts frequently appear in narrative traditions as landscapes where knowledge arises through exposure rather than shelter. Unlike forests, which offer concealment and relational depth, or mountains, which generate transformation through ascent, deserts teach by stripping perception to its essentials. They confront characters with distance, heat, scarcity, and horizon, conditions that narrow attention and heighten awareness. In *Dune* (1965), this relation between exposure and understanding is foundational. Arrakis is not merely a setting but an education in scarcity. Paul Atreides must learn to read the desert: to sense wind shifts, identify dangers in sand movement, walk without rhythm to avoid sandworms, and discipline fear before it overwhelms him. The mantra 'Fear is the mind-killer' is not simply psychological, it is ecological. The desert demands a kind of attention that is calm, precise, and attuned to nuance; panic is lethal because it disrupts perception. In *Yellowstone* (2018–2024), Casey Dutton's retreat into the desert functions as a moment of deliberate withdrawal from social obligation and moral entanglement. Guided by Indigenous spiritual practice, Casey enters an arid, exposed landscape to seek clarity through isolation, fasting, and endurance. Stripped of shelter and distraction, the desert becomes a site where decision-making is deferred until insight emerges through bodily vulnerability and sustained attention. The scene draws on a long cultural association between desert landscapes and revelation, positioning openness, deprivation, and elemental exposure as conditions for ethical reckoning. Rather than offering transcendence or mastery, the desert compels stillness and humility, framing clarity as something that arises through remaining within uncertainty rather than escaping it.

A similar logic unfolds in Paul Bowles' *The Sheltering Sky* (1949), a landmark novel of desert encounter and psychological unravelling. For Bowles' characters, the Sahara is not a backdrop for adventure but a force that dismantles illusion through exposure. As Port and Kit Moresby travel deeper into the desert, its vastness erodes their sense of time, certainty, and social identity. Bowles writes that "the sky hides nothing," capturing the desert's refusal to offer concealment or comfort. The landscape gradually teaches them that the structures that organise ordinary life of routine, narrative, even selfhood don't hold in an environment where the horizon expands endlessly and the sun tolerates no ambiguity. The desert becomes a lesson in existential discipline: one must confront what remains when all external markers fade. Port's illness, displacement, and eventual death are not presented as exceptions but as consequences of failing to understand the desert's terms. Kit's survival depends on learning to move with the landscape's demands rather than resisting them. The desert in Bowles's novel becomes a teacher that strips life to its most elemental forms of attention, endurance, and submission to scale.

Another work that illuminates this desert logic is Michael Ondaatje's *The English Patient* (1992), later adapted into Anthony Minghella's film (1996). In this narrative, the North African desert becomes a landscape that disciplines understanding by removing the familiar structures that sustain identity. The desert's immensity forces characters into a confrontation with space and time on scales far larger than themselves. Ondaatje repeatedly describes the desert as a place where "there is no measure," where distance, horizon, and light distort ordinary perception. This distortion is not simply visual, it transforms how characters interpret memory, loyalty, and truth. For Almásy, the desert becomes an epistemological teacher. As a cartographer, he arrives believing in maps and boundaries, but the environment erodes these assumptions. Lines on paper lose meaning in a landscape that reshapes itself daily through wind and shifting dunes. In the desert, Almásy learns that knowledge is provisional dependent on light, angle, and endurance and that the desert landscape resists being enclosed. Almásy survival hinges not on imposing order but on submitting to the desert's own logic of movement, patience, and orientation. The desert teaches him, in effect, to distrust certainty and to read the world through disciplined attention rather than mastery.

The film adaptation amplifies these themes through visual composition. Long shots of caravans crossing dunes, bodies dwarfed by light and sand, and the quiet interiority of the cave sequences reveal how the desert disciplines perception. The wide horizons refuse narrative closure, and the desert landscape forces the viewer to sit with ambiguity, mirroring Ondaatje's insistence that desert knowledge is fragmentary and unstable. The desert becomes a narrative force that undoes characters as much as it teaches them, reducing their certainties and reconfiguring what they know about love, war, and themselves. Like *Dune* and *The Sheltering Sky*, *The English Patient* shows that desert landscapes generate knowledge not through accumulation but through unmaking. The desert disciplines by requiring stillness, humility, and endurance. Its lessons emerge through exposure, scarcity, and the collapse of conventional orientation. Characters learn not by adding insight but by losing illusions, by realising that the desert offers no mercy, no shelter, and no shortcuts to understanding.

Scarcity, Extraction, and Imperial Ambition

Deserts in narrative landscapes consistently become terrains where scarcity structures power, and where the absence of life-sustaining resources becomes the central axis around which social, political, and ecological relations are organised. Their aridity is not simply atmospheric, it creates a system in which control over water, fuel, minerals, or even mobility becomes synonymous with control over life. In these stories, the desert is not an empty backdrop but a political infrastructure, a landscape that disciplines bodies and legitimises hierarchies through environmental constraint. *Mad Max: Fury Road* (2015) distils this logic into its purest form. Immortan Joe's authority is inseparable from his command of aquifers with water framed as a rationed spectacle, poured in torrents only to be withdrawn as quickly as it appears. Fuel, likewise, becomes a covenant between warlords. Here, the desert's scarcity produces an *economy of desperation*, a world where geography becomes militarised. The desert dictates who survives, who starves, and who rules. Television pushes this logic into late capitalism.

Landman (2024–) portrays the Texas oil fields as contemporary desert landscapes where petro-capitalism thrives on cycles of boom and collapse. Dust, drought, and chemical runoff form a visual rhetoric of depletion, marking the desert as a terrain continually re-inscribed by industrial extraction. In this framing, the desert is neither empty nor inert but actively worked, exposed, and exhausted—becoming simultaneously the host of capital accumulation and the hostage of its environmental consequences. In *There Will Be Blood* (2007), based on Upton Sinclair's novel *Oil!* (1927) is a sweeping social and political narrative set during the early twentieth-century petroleum boom in Southern California. Although not always read as a 'desert novel,' much of its action unfolds in the arid oilfields of California, where dusty plains, sparse settlements, and barren hills form the industrial desert produced by extraction. Sinclair uses this landscape to expose how resource scarcity, speculative ambition, and corporate power reshape both environment and society. In the 2007 film dry land becomes canvas for drilling, speculation, and ecological devastation. Scarcity produces not enlightenment, as in spiritual narratives, but avarice, a frantic search for economic gain through domination of the earth.

War narratives deepen this tension. In *Jarhead* (2005) and *Three Kings* (1999), the Middle Eastern desert is depicted through oil's burnt residue: of strafed dunes, flaming wells, and blackened horizons. These films reveal how desert scarcity is never neutral. The land is framed as valuable only for what lies beneath it, and expendable by the warfare that occurs on it. The scorched sands become a visible index of imperial ambition, territorial desire, and the violent infrastructure of resource extraction. Other films portray deserts not as battlefields but as logistical terrains shaped by global commerce. In *Syriana* (2005), sweeping desert shots frame the geopolitical choreography of oil pipelines, tankers, and multinational negotiations. The desert is transformed into a map of competing claims, each scar of industrial development marking the land as a ledger of extractive intent. Science fiction frequently reimagines desert scarcity as interplanetary resource politics. In *The Expanse* (2011–2022; TV 2015–22), Mars is a mythic desert whose terraforming project becomes a collective ideology with water scarcity and atmosphere shaping civilisation's temperament, discipline, and militaristic ambition. Even *Arrival* (2016), though not set in a literal desert, uses barren Montana plains to evoke the same affective atmosphere: a landscape emptied of distraction in which scarcity of information becomes the central tension. Action cinema reproduces desert logic. *Furious 7* (2015) uses deserts as space where extraction's ultimate fantasy plays out: abundance in the midst of nothing and technological triumph over environmental indifference. It is spectacle built on an assumption of emptiness. Similar themes appear in *Theeb* (2014), where the Arabian desert is depicted with ethnographic precision. Resource scarcity structures social roles with water wells, caravan routes, and tribal alliances becoming lifelines in a world where missteps are fatal. The desert disciplines not through overt violence but through the sheer indifference of landscape. What appears culturally neutral is, in fact, a deeply relational system of survival, trust, and betrayal, all shaped by the scarcity of water, shade, and distance.

Terra Nullius and the Politics of Sacrificial Landscapes

> The doctrine of terra nullius was a lie.
> —*Mabo v Queensland (No 2)*, High Court of Australia (1992)

To treat deserts as deprivation or worse as absence is to risk reproducing the colonial logic of *terra nullius*, the claim that land appearing empty is land available for occupation, extraction, or experimentation. This misrecognition continues to shape how deserts are used, imagined, and narratively mobilised. As Burrows, Holden, and Tynan (2023) demonstrate, the perceived *emptiness* of Maralinga rationalised its selection as a site for British nuclear testing in the 1950 and 1960s. The desert's supposed barrenness became a bureaucratic convenience because the land appeared uninhabited to those in power meaning it could be designated as expendable. Warren (2020) terms this *Maralinga nullius*, a nuclear reiteration of colonial reasoning that cast desert country as an appropriate void for weaponised physics, its ecological, cultural, and ancestral life overwritten by military necessity. Under this framing, the desert becomes both shield and sacrifice. It is a landscape imagined as capable of absorbing contamination, violence, and institutional risk so that other, more valued spaces can remain intact. The act of naming a desert empty performs a double erasure: it denies the presence of people, histories, and nonhuman life, and it conceals the violence of turning a living environment into a receptacle for empire's debris. The politics of sacrifice is always spatial.

Maralinga is not unique. The same logic structured the American nuclear testing programme in Nevada, where the desert was framed as a blank space suitable for atmospheric and underground detonations, so empty, in fact, that mushroom clouds were once marketed as tourist attractions in Las Vegas. The land's Indigenous histories and cosmologies were rendered invisible, replaced by a techno-militarised imaginary that treated the desert as a laboratory. Likewise, the Bikini and Enewetak Atolls, though islands, not deserts, were subjected to similar reasoning: spaces declared null in order to make them available for destruction. Narrative landscapes often reproduce or critique this logic. In Cormac McCarthy's *Blood Meridian* (1985), the desert Southwest becomes a sacrificial ground in which violence is rendered possible by the belief that nothing meaningful occupies the land. McCarthy exposes how emptiness is not a property of the land but a projection of imperial vision. In cinematic dystopias, deserts are frequently the endpoint of civilisation's collapse, a spatial metaphor for what remains after the world has exhausted its capacity for renewal. In *The Matrix* (1999), Morpheus's phrase 'Welcome to the desert of the real' names not a literal terrain but a condition: the world left behind after illusion collapses. The desert becomes emblematic of exposure, the stripped-down residue of a world hollowed by systems of bureaucratic and capitalist control. Its imagery draws directly on landscapes of waste, ruin, and technological aftermath, the same terrains empire designates as expendable.

In Josh Fox's *Gasland* (2010), fracking sites scattered across the dry American interior are similarly framed as expendable terrains. They are landscapes designated to absorb the environmental and social costs of national energy appetites. Fox's camera lingers on dusty plains punctured by drill pads, pipelines, wastewater ponds, and flare stacks; an industrial architecture dropped onto regions narratively coded as empty or marginal. The film repeatedly exposes how these arid geographies are treated as *sacrifice zones*, where groundwater contamination, air pollution, chemical runoff, and community displacement are rendered politically acceptable precisely because the land is imagined as peripheral. Residents ignite tap water, breathe in drilling fumes, and watch their wells sour, yet the surrounding land continues to be presented as spacious, uninhabited, and available for exploitation. *Gasland* reveals the paradox at the heart of extractive modernity and petro-culture: the very regions considered barren are those most burdened with environmental risk (Nixon 2011). The desert or semi-arid interior becomes the space where industrial excess is allowed to accumulate, its ecological and human costs invisibilised by a national narrative that casts dry landscapes as durable, empty, or insignificant. What these examples reveal is that deserts are rarely understood on their own terms. They are routinely constructed as buffers, testing grounds, or extraction sites where the consequences of corporate, political or imperial ambitions can be displaced. Desert landscapes, hence, are imagined as strong enough, empty enough, or insignificant enough to absorb impacts that would be unacceptable elsewhere.

Island as Epistemic Capture: Universalism and Cultural Appropriation

Islands in the stream
That is what we are
No one in between
How can we be wrong?
Sail away with me
To another world

The island occupies a distinct position within the long tradition of mythical landscapes. Islands are a microcosm where new configurations of life, knowledge, and relation can be created or discovered. From *Atlantis* in Plato's *Timaeus* and *Critias* to Avalon in medieval legend, Tolkien's Tol Eressëa, and the Laputas of Swift and Miyazaki, the island recurs as a bounded world where distance from the ordinary permits alternative orders of life and thought. *Atlantis* joins knowledge with pride; Avalon and Tol Eressëa offer rest beyond mortality; and both floating Laputas expose the fragility of intellect detached from the world; Swift's island mocks the sterility of reason turned inward; and Miyazaki's drifting city closes the loop. Miyazaki reworks Atlantis as a sky-borne parable: a technological paradise undone by its own mastery, leaving only remnants that instruct later generations. Their modern counterparts maintain the same structural rule: enclosure regulates relations, and distance enables experiment. Hence, the island is both an epistemic instrument and a colonial laboratory. Islands as evolutionary and cultural laboratories rely on their ecology to reinforce this narrative function.

Gloria Pungetti considers islands as laboratories for both evolutionary and cultural processes (2016), drawing on Darwin's observation that isolation produces differentiation. On the Galápagos ecological boundaries shapes form, behaviour, and possibility. The same mechanism applies to cultural space. Narrative islands become mental boundaries that permit experiment by excluding interference. Ian Kinane builds on this in *Theorising Literary Islands* (2016) by examining the island as a persistent literary trope shaped by colonial, imperial, and neoliberal imaginaries. Building on Robinsonade traditions, Kinane argues that islands are repeatedly framed as bounded, exceptional spaces that invite experimentation, control, and reinvention, often under the fiction of emptiness or isolation. He shows how contemporary island narratives continue to reproduce and critique these logics, positioning islands as laboratories for sovereignty, survival, and ethical testing. Crucially, Kinane emphasises that islands are never truly isolated or neutral, rather they are relational spaces embedded within global networks of power, capital, and mobility, making them key sites for interrogating myths of autonomy, *terra nullius*, and spatial innocence.

The Island as Controlled World: Isolation as Method

Across literary and cinematic history, the island functions as a technical device, a bounded world where complexity can be rendered legible. Isolation becomes a method for transforming existence into experiment. In *Robinson Crusoe* (Defoe, 1719), the sea isolates the protagonist within a controlled field of reconstruction. Crusoe's island becomes an instrument for reorganising labour, faith, and governance. Sowing grain, building shelter, domesticating goats, translates bare survival into a lesson in providence and property. The island performs epistemic work, it frames the world small enough that order seems graspable, mastery appears virtuous, and the colonial imagination is naturalised. Nearly two centuries later, *The Island of Dr Moreau* (Wells, 1896) restages this insular experiment through biology. The shoreline becomes a membrane between species where surgical transformation replaces divine providence as the architect of hierarchy. The island's isolation legitimises Moreau's vivisections because no external system of justice, law, or community can intervene. His laboratory depends on spatial containment, without the island, the experiment would collapse under moral scrutiny.

By the late twentieth century, a new Promethean figure industrialises this rationality. In *Jurassic Park* (1990, 1993), the island becomes a corporate research compound where containment systems, visitor pathways, and experimental protocols merge science with spectacle. The isolated and exiled Moreau becomes the entrepreneurial Hammond, who transforms exclusion into enterprise. In *Jurassic World* (2015), the island evolves into a fully operational theme park, its boundaries both protecting the mainland from the creatures and enclosing them within an engineered ecosystem. *The Wilds* (2020–22) presents a clear contemporary uses of the island as a Promethean-controlled world. A group of teenage girls survive what they believe is a plane crash, only to discover that their isolation on a remote island is the product of a covert social experiment designed to test alternatives to patriarchal social structures. The island becomes a laboratory of behaviour, a place where the absence of external society allows researchers to observe the girls' emotional, social, and moral development in conditions of scarcity and stress. The shoreline marks the limit of their world, creating a boundary that intensifies the consequences of every decision. Without outside interference, group dynamics surface with uncomfortable clarity: alliances form and fracture; power circulates unpredictably; and the island's environment pressures the girls into revealing their resilience and vulnerabilities. The experiment depends entirely on isolation, only a controlled world can produce the behavioural data the researchers seek. The island's apparent wildness conceals total surveillance. Hidden cameras track movement, microphones capture private conversations, and supply drops subtly manipulate the group's chances of survival. The island setting thus becomes a double enclosure because the girls are trapped within the island's geography, and the island itself is trapped within an institutional apparatus that both shapes and interprets their lives. In this sense, *The Wilds* recasts the Robinsonade island as a space where identity is not discovered but constructed, even engineered, through isolation.

M. Night Shyamalan's *Old* (2021) offers a similarly unsettling version of the island-as-experiment by using a secluded beach as a biomedical testing ground. The beach, accessible only through a narrow rock passage, functions as an island within an island: once entered, its boundaries become impenetrable. Time accelerates dramatically within this enclosed zone, transforming hours into decades and turning the beach into a living laboratory where lifespans compress into narrative arcs. This controlled environment is engineered by a pharmaceutical company using the beach's unique temporal anomaly to conduct rapid clinical trials on human subjects. The island's isolation enables what would otherwise be impossible, observing the long-term effects of drugs within a single day. As with classic island laboratory narratives, the experiment depends on removing its subjects from the ethical, moral, legal, and social constraints of the wider world. The beach becomes a landscape that narrates death, aging, disease, and fear at accelerated speed. Everybody becomes a data point and every relationship is reshaped by the compression of life into a few desperate hours. What appears at first as an idyllic tropical cove, reveals itself as an instrument of extraction, where the value of human life is subordinated to the production of corporate medical knowledge. But the experiment works only because the beach is isolated spatially, temporally, and ethically. The island amplifies vulnerability by erasing escape.

Heterotopian Islands: Spaces That Mirror and Distort the World

Michel Foucault describes heterotopias as real places that gather together incompatible meanings. They are sites that reflect and distort the world simultaneously (Foucault, 1986). Islands often take this form in narratives because their isolation allows contradictory orders to coexist within a single bounded terrain. They provide distance from the everyday world, yet they reproduce its structures in intensified, concentrated form. The island becomes a mirror that shows the world not as it imagines itself to be, but as it actually functions beneath the surface. Danny Boyle's *The Beach* (2000) constructs a clear cinematic example of the island as heterotopia. The community that forms on the island believes it has left modernity behind, seeking purity, freedom, and escape. Yet its labour routines, surveillance practices, sexual politics, and systems of punishment replicate the very world it rejects. In Foucault's terms, the island becomes a counter-site, a place that exposes the illusions of utopian withdrawal by replaying the dynamics of control, secrecy, and discipline in a closed environment. Even the film's production reinforces this heterotopian logic: the pristine sand and turquoise water presented to audiences were constructed through ecological manipulation, with imported sand reshaping an existing landscape to fit Western fantasies of paradise. The island is both symbol and product of the illusion it stages, a heterotopia built at the level of both story and world.

Foucault's example of the ship as heterotopia; a floating piece of space, self-contained yet exposed to the infinity of the sea illuminates *Life of Pi* (2012). Pi's lifeboat becomes a mobile microcosm governed by strict rules of coexistence and ritual, its fragile order constantly negotiated between human and animal. The boat is not a site of freedom but of necessary governance, a heterotopia where survival depends on maintaining structure in a world stripped of certainty. The carnivorous island encountered later amplifies this doubleness. Island can be a sanctuary and a trap: a feminised, consuming landscape that nurtures and annihilates and a space where binaries such as consumer/consumed and coloniser/colonised collapse and a dream-projection of repressed violence. Its fruits, sweet by day, lethal by night, literalise the cost of refuge. The island offers safety only at the price of entrapment, revealing the heterotopian condition in its most extreme form: a place where contradictory states, paradise and peril are inseparable.

The Truman Show (1998) and *Shutter Island* (2010): Carceral Heterotopian Islands of Surveillance, Simulation, and Control

The Truman Show (1998) constructs one of the most sophisticated heterotopian islands in contemporary cinema. Seahaven is an island, encircled by water and psychologically bounded by Truman's cultivated fear of the sea. Yet its insularity is more than geographic. It is a fabricated total environment, a world built to mirror the structures of suburban America while simultaneously distorting them into instruments of surveillance and control. The island is a heterotopia, it is a real place that gathers contradictory meanings, a site where the familiar becomes *uncan*ny through the intensification of normative order (Foucault, 1986).

Seahaven presents itself as perfect ordinariness of sunlit streets, congenial neighbours, and spotless houses but this surface uniformity renders it weird and eerie. Mark Fisher distinguishes between the weird and the eerie as two modes through which culture registers disturbances in reality. The weird is defined by the presence of something that does not belong as an intrusion that disrupts established ontologies while the eerie is marked by absence, silence, or the sense that something expected is missing (Fisher, 2016). Truman's feelings of the weird and the errieresonate in the film with his emerging consciousness of things that don't belong, their intrusion, and that something is missing. There are key scenes in which Truman's wife, abruptly slips into a glossy product endorsement while standing in their kitchen. Holding up items such as cocoa or kitchen appliances, she delivers advertising copy directly to Truman, her tone flat and rehearsed, as if addressing an invisible audience rather than her husband. Truman's response is confusion that quickly turns to distress and anger. He interrupts her, asking "Who are you talking to?" and becomes visibly shaken as the emotional reality of the moment collapses into commercial performance. Other moments occur when Truman drives to the outskirts of Seahaven and encounters repeated obstacles (traffic jams, forest fires, nuclear alerts) that prevent him from leaving. These barriers draw attention not to what is present, but to what is absent: genuine routes out, unscripted risk, and a world that exceeds him. The environment behaves as if agency has been removed, producing the unsettling realisation that the landscape exists only insofar as it serves him. More broadly, the town of Seahaven is composed through symmetrical framing, pastel colour palettes, and controlled lighting that eliminate ambiguity and depth. This hyper-legibility creates unease: nothing is hidden, yet something is missing, namely, autonomy, interiority, and unpredictability. In Fisher's terms, the *mise-en-scène* produces the eerie not through excess or spectacle, but through the withdrawal of agency, making the world appear intact while revealing its fundamental emptiness.

Returning to Foucault, surveillance is the island's primary organising principle. Thousands of cameras are embedded into the fabric of Seahaven, in lampposts, car mirrors, radios, and household objects producing a network of observation that maps Truman's movements continuously. In Foucault's logic, Seahaven becomes a *panoptic heterotopia*, a space where the subject is fully visible while the mechanisms watching him remain hidden. The dome itself functions as both sky and apparatus, an architectural interpretation of Bentham's panopticon scaled to the size of a town. Truman's life is recorded not for his benefit but for the consumption of others, transforming the island into a heterotopia of exhibition, a museum of one man's existence, curated in real time. The island's fakeness intensifies its heterotopian function. Seahaven simulates a coherent world by excluding all threats to its stability. No poverty, illness, conflict, or unpredictability is permitted, this also reflects ideas of the *eerie* and the *weird* (Fisher, 2016). Everything is managed to preserve the illusion of seamless normality. In this sense, the island also operates as a *compensatory heterotopia*, a place that attempts to perfect what the outside world cannot. But beneath this perfection lies a disciplinary order in which Truman's emotions, desires, routines, and relationships are regulated through spatial design. His fear of water is implanted, his friendships are cast and even his marriage is orchestrated. Seahaven is a space where autonomy is performed rather than lived. The island's boundaries enforce this illusion. The surrounding ocean does not simply demarcate physical space, it produces Truman's psychological enclosure. When he tries to leave, the island responds like a living immune system, generating traffic jams, hazardous weather, and scripted emergencies. The island defends itself. It does so because its success as heterotopia depends on Truman's containment, escape would collapse the carefully managed equilibrium between simulation and belief.

Moments of rupture within Seahaven also reveal its *heterotopian instability*. A stage light falls from the "sky." A radio channel accidentally intercepts production cues. An elevator opens onto a backstage catering area. The moment Truman touches the dome's boundary and finds the staircase hidden in the horizon, the heterotopia's secret becomes visible: the world he knows is not the world as it is, but the world as it has been made for him. Thus, *The Truman Show* presents the island as a heterotopia of surveillance, simulation, and disciplinary control. Seahaven is an island that mirrors the social world but intensifies its governing structures into a closed-circuit system. It reveals the mechanisms of consumer culture, the performance of everyday life, and the architecture of mediated experience by stripping away the possibility of unstructured existence. The island does not free Truman from society; it condenses society into an apparatus that captures him. His final gesture, the decision to step through the door at the back of the sky marks *heterotopian collapse*, the moment when the bounded world can no longer sustain its fiction. *The Truman Show* demonstrates how the island, as narrative landscape, becomes a site where power is not only displayed but perfected. Seahaven is not utopia, nor dystopia, but a heterotopia that reveals the world by staging its most controlled and curated version.

Shutter Island (2010) presents a different but equally potent form of the heterotopian island, one structured around psychiatric power, carceral surveillance, and the orchestration of reality. Ashecliffe Hospital, perched on the storm-lashed island, is a closed world that mirrors the architectures of surveillance and discipline identified by Foucault in his analyses of prisons and asylums (Foucault, 1961, 1977). The island's isolation is absolute with its steep cliffs and violent seas forming an unbreachable perimeter that functions both as geographic boundary and psychological containment. This enforced separation allows the institution to reorganise perception, turning the island into a laboratory of behaviour where the distinction between treatment and control and truth and fabrication begins to collapse. It is at once hospital and prison, sanctuary and threat, experiment and punishment. Every building on the island (Ward A, Ward C, the lighthouse) embodies a different modality of surveillance and coercion. Foucault describes heterotopias as spaces where incompatible functions coexist (Foucault, 1986) and Ashecliffe enacts this precisely: rehabilitation is inseparable from discipline, care from violence, and truth from manipulation. The institutional façade of healing conceals a regime of total observation in which doctors, orderlies, and guards monitor not only the patients but also the protagonist, Teddy Daniels. His investigation is not a neutral inquiry but part of the island's design, an orchestrated sequence meant to expose and reshape his identity.

Unlike Seahaven's polished suburban veneer, Ashecliffe embraces the aesthetic of decay with crumbling walls, dim corridors, barred windows, and rocky cliffs. Yet this deterioration is part of its heterotopian power. The island's darkness, storms, and labyrinthine structures destabilise Teddy's sense of orientation, making him increasingly susceptible to the institution's narrative of his own madness. Every escape attempt, like Truman's attempted crossings of Seahaven's boundaries, is met with an intensification of spatial constraints of electrified fences, locked gates, and disciplined orderlies resulting in the physical impossibility of escaping the island. The island becomes an epistemic trap: a space where every path leads back to the institution's framing of the subject, and where madness becomes indistinguishable from environment. Like Seahaven, Shutter Island reveals how the island form enables total control. Isolation allows the institution to manipulate perception, determine truth claims, and construct a reality in which the subject's resistance is both anticipated and absorbed. The island mirrors the world's disciplinary structures of medicalisation, confinement and surveillance but amplifies them until they become luminous. In this way, *Shutter Island* stands as a heterotopian island where enclosure, simulation, and the choreography of psychological power produce a world that exposes more than it contains.

Caves as Erotic and Gendered Landscapes: Descent, Relation, and Interior Knowledge

I was not afraid in the cave.
—*The Descent* (2005)

Across religious, mythic, and psychological traditions, the cave appears as a landscape of inward movement, an environment where descent becomes a return to the substance from which consciousness arises. Jung describes this inward motion as the psyche's retreat into its own originary ground, a re-entry into the *womb of thought* where awareness rediscovers the material matrix of its formation (Jung, 1964). Eliade similarly understands the cavern as the earth's breathing chamber, a site where death and renewal coincide, and where what decays becomes the seed of future life (Eliade, 1959). Plato disrupts this continuity. In *The Republic* (1997, 514a–518b), the cave ceases to nourish and becomes instead the emblem of illusion; truth is relocated outside, in a realm of pure form, while the subterranean space is left behind as a site of ignorance. From this inversion descends a long intellectual lineage that equates the feminine, the body, and the earth with opacity. The philosopher's climb toward light becomes the first philosophical act of separation, positioning transcendence against the material world that once sustained it. Here, the cave becomes a narrative landscape that stages the first fracture between body and mind, world and thought.

Feminist thinkers repeatedly return to the cave to expose the epistemological and ontological consequences of Plato's foundational separation between light and darkness, reason and matter, ascent and enclosure. Where the Platonic cave allegory frames subterranean space as a site of illusion to be escaped, feminist re-readings interrogate what is lost when enclosure is cast solely as deprivation. Çimen (2025) identifies a decisive shift in Porphyry's *Cave of the Nymphs*, where the womb's generative power is re-coded as cosmic mechanism. In this reworking, the cave is transformed from a living matrix into a vaulted architecture of abstract order, prefiguring later rationalist and militarised enclosures that privilege control, hierarchy, and masculine reason over material continuity and care. This sense reverberates in the modern bunker, a structure that inherits the cave's enclosure while stripping it of generativity. The bunker promises protection through isolation, survival through withdrawal, and knowledge through command-and-control visibility. Feminist critiques expose how such architectures convert enclosure into a technology of fear, discipline, and exclusion, severing the cave from its associations with gestation, latency, and emergence. Against this, prehistoric and Buddhist cosmologies in which darkness remains productive rather than threatening are animated. In these traditions, the womb-world affirms that containment can sustain creativity without domination, and that not all protection requires fortification or violence. Darkness functions as a condition of emergence rather than an absence to be overcome. Read alongside bunker imaginaries, this distinction becomes politically charged. Where the bunker forecloses relation in the name of security, the cave-as-womb cultivates inwardness without severance, enabling change without retreat from the world. These feminist re-readings recover what Platonic ascent and bunker logic alike erase, the material continuity between birth, thought, and environment. In this reconceptualisation, subterranean space is no longer the antithesis of knowledge but its condition of possibility. The cave is reclaimed as a landscape of immanence, where enclosure supports becoming rather than domination. Feminist cave thinking thus exposes how the politics of enclosure hinge not on whether space is closed, but on whether it remains relational, porous, and oriented toward emergence rather than control.

Subterranean Narratives: Caves in Literature, Film, and Cultural Memory

In popular culture, caves repeatedly function as charged interior landscapes where knowledge is produced through descent, enclosure, and bodily vulnerability rather than mastery or transcendence. Across horror, fantasy, and science fiction, the cave emerges as a feminised and eroticised space: dark, generative, and resistant to visual certainty. *The Descent* (2005) is a British horror film that treats the cave as a driving landscape narrative rather than a neutral backdrop. The plot follows an all-female group of experienced cavers who enter an unmapped cave system, only to become trapped after a collapse seals their exit. As darkness deepens and resources diminish, the women confront both the hostile environment and predatory subterranean creatures, forcing a shift from exploration to survival. Working within the survival-horror genre, the film mobilises claustrophobia, sensory deprivation, and bodily proximity to generate fear, while the cave's spatial constraints reorganise authority, intimacy, and perception. With vision repeatedly compromised, knowledge is produced through touch, breath, sound, and pain, positioning the cave as an active agent that shapes character, conflict, and outcome, foregrounding a gendered, embodied epistemology forged through descent.

A comparable treatment appears in *The Clan of the Cave Bear* (1986) where the cave functions as a landscape of gendered power and exclusion rather than mere shelter. It is a prehistoric drama film based on the 1980 novel of the same name by Jean M. Auel. Set in Ice Age Europe, it portrays a Cro-Magnon girl raised by a Neanderthal tribe, exploring themes of survival, identity, and cultural clash. Ayla's expulsion from the Neanderthal clan marks a prehistoric preface to feminist rebellion where the cavern ceases to operate as a site of enforced silence and instead becomes a dwelling through which resistance, adaptation, and alternative forms of knowledge are sustained. Within the film's landscape narrative, the cave mediates survival, memory, and social order, revealing how enclosure can simultaneously discipline and protect, and how subterranean space enables forms of becoming that exceed patriarchal control.

Susan Glaspell's play *The Verge* (1921) replays this pattern in a greenhouse setting, a glass cave, where confinement and emergence coexist. It is a key work of early modernist and feminist drama that explores themes of individuality, creativity, and societal constraint through expressionist form and language. Like the cave, the play becomes a site where interiority enables transformation rather than silence. This interior logic extends even into ostensibly open landscapes. Weatherby and Vidon's (2018) digital cartographies of female cavers and climbers reveal how wilderness spaces long mythologised as masculine domains or 'man caves' retain an interior pulse shaped by gendered experience. Through mapping, narration, and embodied practice, these climbers re-inscribe vertical and exposed terrains as spaces of relational dwelling rather than conquest, demonstrating that enclosure and interiority are not defined by walls alone but by how space is inhabited, mediated, and made meaningful.

Erotic and transformative patterns are also evident in *The Company of Wolves* (1984), where caves and forest interiors operate as thresholds for sexual awakening and self-recognition. Descent signifies not loss but transition: a movement into spaces where desire and danger coexist, and where identity is reshaped rather than stabilised. Science fiction extends this pattern by reworking the cave as an origin space that is simultaneously alluring and unsettling. In *Alien* (1979) and *Annihilation* (2018), cavernous interiors, whether biological, geological, or extraterrestrial, function as womb-like environments that generate new forms of life and knowledge, often at the cost of bodily integrity. Simone de Beauvoir provides the conceptual architecture for understanding how gender shapes the symbolism of subterranean space. She defines patriarchal ontology through the opposition of transcendence and immanence: man projects himself outward toward projects and futures, while woman remains confined to cycles of maintenance and reproduction (de Beauvoir, 1956). The cave spatialises this dichotomy. Its darkness becomes aligned with immanence, the feminised interior that patriarchal thought seeks to escape. In this reading, Plato's ascent from the cave becomes the symbolic origin of exclusion: knowledge begins only when the philosopher leaves the womb. Luce Irigaray re-enters this forgotten interior to expose its erasure. In *Speculum of the Other Woman*, Irigaray shows how the ascent relies upon suppressing the maternal body that makes birth, and thus philosophy, possible (Irigaray, 1985). Julia Kristeva identifies this interior as the *chora*, a pre-symbolic, pulsating field that precedes language and sustains symbolic order even while being excluded from it (Kristeva, 1984). Feminist reinterpretations make visible the textures of the cave that philosophy once rendered invisible. *The Descent* shows how screams within darkness articulate an embodied grammar of endurance, turning affect into epistemology.

These mythical landscapes of mountains, forests, deserts, islands, and caves reveal how deeply human cultures have relied on the lanscpaes to think through questions of memory, transformation, relation, and meaning. Each terrain operates not only as symbolic form but as narrative method, shaping how stories unfold and how consciousness organises experience. They anchor cosmologies, challenge perception, and offer imaginative frameworks for negotiating identity, power, and belonging. Yet mythical landscapes do more than illuminate what cultures revere, they also cast shadows over what cultures exploit. The same spatial logics that produce revelation can also produce ruin. As the next chapter turns to toxic landscapes where we shift from sacred terrain to polluted ground, examining how industrial modernity reshapes the earth into zones of harm, sacrifice, and slow violence. If mythical landscapes teach us how humans have sought meaning in the world, toxic landscapes reveal how meaning breaks down when landscapes become sites of contamination, materials carrying histories that can't be forgotten, and futures that can't be ignored.

References

Burrows, L., Holden, D., & Tynan, E. (2023). Untangling maralinga: Spatial and temporal complexities of Australia's atomic anthropocene. *Journal of Australian Studies, 47*(3), 515–530. https://doi.org/10.1080/14443058.2023.2199757

Casey, E. S. (1997). *The fate of place: A philosophical history*. University of California Press.

Chakrabarty, D. (2021). *The climate of history in a planetary age*. University of Chicago Press.

Eliade, M. (1959). *The sacred and the profane: The nature of religion* (W. R. Trask, Trans.). Harcourt, Brace.

Fisher, M. (2016). *The weird and the eerie*. Repeater Books.

Foucault, M. (1961). *Madness and civilization: A history of insanity in the age of reason* (R. Howard, Trans.). Pantheon.

Foucault, M. (1977). *Discipline and punish: The birth of the prison* (A. Sheridan, Trans.). Pantheon.

Foucault, M. (1986). Of other spaces. *Diacritics, 16*(1), 22–27. https://doi.org/10.2307/464648

Haraway, D. J. (2016). *Staying with the trouble: Making Kin in the Chthulucene*. Duke University Press.

Harvey, D. (2006). *Spaces of global capitalism: Towards a theory of uneven geographical development*. Verso.

Hillman, J. (1975). *Re-visioning psychology*. Harper & Row.

Ingold, T. (1993). The temporality of the landscape. *World Archaeology, 25*(2), 152–174. https://doi.org/10.1080/00438243.1993.9980235

Ingold, T. (2000). *The perception of the environment: Essays on livelihood, dwelling and skill*. Routledge.

Irigaray, L. (1985). *Speculum of the other woman* (G. C. Gill, Trans.). Cornell University Press.

Jameson, F. (1981). *The political unconscious: Narrative as a socially symbolic act*. Cornell University Press.

Jung, C. G. (1964). *Man and his symbols*. Aldus Books.
Kant, I. (1790). *Critique of the power of judgment (Kritik der Urteilskraft).*
Kimmerer, R. W. (2013). *Braiding sweetgrass: Indigenous wisdom, scientific knowledge, and the teachings of plants*. Milkweed Editions.
Kristeva, J. (1984). *Revolution in poetic language* (M. Waller, Trans.). Columbia University Press.
Marcuse, H. (1955). *Eros and civilization: A philosophical inquiry into freud*. Beacon Press.
Næss, A. (1973). The shallow and the deep, long-range ecology movement: A summary. *Inquiry, 16*(1–4), 95–100. https://doi.org/10.1080/00201747308601682
Næss, A. (1989). *Ecology, community and lifestyle: Outline of an ecosophy* (D. Rothenberg, Trans.). Cambridge University Press.
Nagel, T. (1974). What is it like to be a bat? *The Philosophical Review, 83*(4), 435–450. https://doi.org/10.2307/2183914
Nixon, R. (2011). *Slow violence and the environmentalism of the poor*. Harvard University Press.
Pungetti, G. (2016). *Island landscapes: An expression of European culture*. Routledge.
Rose, G. (1993). *Feminism and geography: The limits of geographical knowledge*. Polity Press.
Schama, S. (1995). *Landscape and Memory*. Alfred A. Knopf.
Simard, S. W., Beiler, K. J., Bingham, M. A., Deslippe, J. R., Philip, L. J., & Teste, F. P. (2012). Mycorrhizal networks: Mechanisms, ecology and modelling. *Fungal Biology Reviews, 26*(1), 39–60. https://doi.org/10.1016/j.fbr.2012.01.001
Simard, S. W., Perry, D. A., Jones, M. D., Myrold, D. D., Durall, D. M., & Molina, R. (1997). Net transfer of carbon between ectomycorrhizal tree species in the field. *Nature, 388*, 579–582. https://doi.org/10.1038/41557
Tuan, Y.-F. (1990). *Topophilia: A study of environmental perception, attitudes, and values*. Columbia University Press.
Vico, G. (1744). *The new science (Scienza Nuova)*. Giambattista Vico (Stanford Encyclopedia of Philosophy).
von Uexküll, J. (1909). Umwelt und innenwelt der Tiere. Springer.
von Uexküll, J. (2010). A foray into the worlds of animals and humans: With a theory of meaning (J. D. O'Neil, Trans.). University of Minnesota Press.
Warren, J. G. (2020). Apprehending the slow violence of nuclear colonialism: Art and maralinga. In A. Llewellyn (Ed.), *Australia's Arc of Fire: Indigenous Activism, the Environment and the Postcolonial State* (pp. 131–152). UTS ePRESS.

Weatherby, T. G., & Vidon, E. S. (2018). Delegitimizing wilderness as the man cave: The role of social media in female wilderness empowerment. *Tourist Studies, 18*(3), 332–352. https://doi.org/10.1177/1468797618771691

Wylie, J. (2007). *Landscape*. Routledge.

CHAPTER 5

Contaminated Grounds and the Ecologies of Toxic Modernity

Abstract Contaminated ground is considered a central condition of modernity rather than an exceptional environmental problem. Toxic landscapes emerge through the entangled histories of industrialism, colonialism, and capitalist extraction, which inscribe harm into soil, water, air, infrastructure, and bodies. These sites function as historical archives: they store residues of development, state violence, and environmental neglect, and they transmit injury across generations through heavy metals, radioactive isotopes, pesticide drift, petrochemical by-products, and infrastructural abandonment. Because contamination is often diffuse, delayed, and difficult to see, its violence unfolds as slow violence, an attritional harm that accumulates through sediments, groundwater, cellular damage, and inherited vulnerability rather than erupting as a single event. The distribution of toxicity is therefore political: hazardous industries and waste infrastructures concentrate on Indigenous lands and marginalised communities, making exposure a mode of governance organised through sacrifice, abandonment, and managed precarity. Toxicity is analysed through three linked dimensions: atmosphere, residue, and time. Atmosphere names the sensorial and affective experience of living with contaminated air, where breath becomes precarious and unease precedes proof. Residue tracks the stubborn material persistence of harm in tailings, ruins, domestic dust, and poisoned soils, producing damaged ecologies that endure long after industries depart. Time captures latency and afterlife: illness and loss arriving decades later, remediation stretched across generations, and

J. Coghlan and H. Nolan, *Landscapes That Speak*,
https://doi.org/10.1007/978-3-032-21757-8_5

closure repeatedly deferred. Screen cultures are used to show how toxicity becomes perceptible through form: colour, haze, soundscapes, silence, and sensory disturbance render contamination as lived condition as much as measurable fact. Toxic landscapes thus appear as both evidence and battleground, where the afterlives of modernity's promises of progress are felt, contested, and borne.

Keywords Contamination · Toxicity · Sacrifice · Latency · Residue · Atmosphere · Slow violence

Contaminated ground is a defining condition of modernity, produced through the intertwined forces of industrialism, colonialism, and capitalist extraction that have inscribed harm into soils, waters, and bodies. Toxic landscapes are not merely polluted sites but historical archives, storing the accumulated residues of development, state violence, and environmental neglect. Whether chemically saturated, radioactively unstable, agriculturally exhausted, or infrastructurally abandoned, contaminated terrains carry the lived histories of harm across generations and geographies. Contamination operates both materially and symbolically: as chemical residue, political aftermath, sensory disturbance, and affective atmosphere that reshapes how life is experienced, remembered, and endured. These landscapes unsettle the boundary between ecology and history, as heavy metals, radioactive isotopes, pesticide drift, and postcolonial detritus persist across lifetimes, embedding themselves in everyday existence. Their impacts unfold slowly and unevenly, producing *slow violence*, forms of delayed destruction that accumulate through groundwater, sediment, cellular damage, and inherited vulnerability rather than erupting as discrete events (Nixon, 2011). Crucially, contamination is unevenly distributed. It follows the spatial logics of colonisation, racial capitalism, and state abandonment, concentrating hazardous industries and waste infrastructures on Indigenous lands and low-income communities designated as sacrificial to extraction and national development. Toxicity thus functions as a mode of governance, organising land and life through managed exposure and environmental precarity. In this context, contaminated landscapes are not passive backdrops but active terrains of political struggle, where bodies and environments bear witness to the enduring afterlives of state-sanctioned harm.

Toxicity as Atmosphere, Residue, Time

Contamination operates through three interwoven dimensions: atmosphere, residue, and time. As atmosphere, it is sensorially embodied in air thick with smog or odours of oil or sulphur. As residue, it embeds industrial and settler histories into soil and water, producing ecologies that manifest as corporeal and poisonous legacies, saturated with lived experience and intergenerational trauma. As time toxicity disrupts causality, unfolding through delay and repetition with illnesses manifesting decades after exposure, clean-ups stretching across generations and with remediation perpetually deferred. These dimensions each articulate a temporality without closure. The *slow violence* of toxic contamination weakens legal, political, and moral structures that rely on identifiable perpetrators, undermining the pursuit of justice and the possibility of fair human and environmental redress; if this is even possible.

Pollution does not occur randomly; it follows the cartographies of colonisation and capital. Hazardous industries, mining operations, and waste disposal sites are overwhelmingly situated on Indigenous, impoverished, or otherwise marginalised lands. Contamination is not only a residue of industrial extraction but an ongoing structure of governance that renders some lives expendable and some landscapes perpetually unliveable. Toxicity is both a condition and a metaphor for modernity's unfinished violences. It transforms landscapes into archives, bodies into evidence, and futures into uncertain inheritances. The persistence of radioactive isotopes, plastics, and heavy metals reveals the deep entanglement between human temporality and geological time. This entanglement defines the Anthropocene: a moment when the long durations of planetary processes converge with human histories of exploitation and inequality. Contaminated landscapes are the material form of that convergence. They compel us to recognise that environmental crisis is not a historical break but the predictable outcome of corporate extraction, governmental sanction, and the relentless logics of capitalist greed. To study contaminated ground, then, is to trace the afterlives of modernity's promises of progress, prosperity and control and to see in their decay the human and environmental cost of their fulfilment.

To speak of contaminated ground is to invoke more than the material presence of pollutants or chemical waste. Toxicity inhabits space and time and binds bodies, environments, and histories in uneven ways. Contaminated ground does not only designate the measurable thresholds of poisoned air, soil, or water but refers to an expanded field of experience in which contamination reshapes identity and memory and persists across durations that exceed human temporalities. They are lived as slow encroachments and as atmospheres of uneases. They are affective and corporeal residues that insist long after their supposed neutralisation. Rob Nixon's account of *slow violence* offers a crucial framework for grasping how contamination harms incrementally. It is worth citing Nixon account in some detail. He asserts that slow violence is

> … a violence of delayed destruction that is dispersed across time and space, an attritional violence that is typically not viewed as violence at all. Violence is customarily conceived as an event or action that is immediate in time, explosive and spectacular in space, and as erupting into instant sensational visibility … violence that is neither spectacular nor instantaneous, but rather incremental and accretive, its calamitous repercussions playing out across a range of temporal scales … Climate change, the thawing cryosphere, toxic drift, biomagnification, deforestation, the radioactive aftermaths of wars, acidifying oceans, and a host of other slowly unfolding environmental catastrophes … [Slow violence is] the long dyings—the staggered and staggeringly discounted casualties, both human and ecological that result from war's toxic aftermaths of climate change. (Nixon, 2011, 2–3)

Unlike spectacular disasters of explosions, floods, and earthquakes, which cohere into recognisable events demanding urgent attention, toxic violence rarely announces itself. It is diffuse, attritional, and temporally stretched. Toxins seep silently into water tables and soil, accumulating in cellular structures, and producing forms of harm whose temporality is belated, often only perceptible years or decades later. This latency generates epistemological and political difficulties because its violence resists the drama of instant catastrophe. It eludes media spectacle, judicial recognition, and collective memory. Slow violence is neither spectacular nor instantaneous but rather incremental and accretive, its calamitous repercussions playing out across a range of temporal scales. What makes slow violence especially pertinent to contaminated is its dual ontology. On the one hand, contamination is insistently material: it leaves chemical residues, scars on both landscapes and communities, and results in corporeal mutations both physical and emotional. On the other hand, it is spectral, operating through invisibility, latency, and absence. A contaminated field may look pristine yet conceal carcinogenic soil, water may appear clear while carrying unseeable poisons. For Nixon, such violence is typically not viewed as violence at all precisely because it resists being seen, dramatised, or spectacularised. Toxic landscapes thus stage a paradox of presence and absence: they are hyper-material in their density of pollutants and harm yet spectral in the way their effects are diffuse and molecularly hidden. This paradox destabilises conventional distinctions between event and aftermath. In the case of the 1984 Bhopal disaster,[1] Fukushima[2] and other recent industrial and extractive 'events'[3] there is are toxic afterlives. Even after remediation efforts, residues persist, contaminating soil, water, and bodies. As a result, toxicity becomes not a singular occurrence but a temporally open process, one that continually reasserts itself against narratives of closure or recovery. By reframing violence as slow, Nixon foregrounds the ways toxicity reconfigures landscapes as temporal, political, corporeal and affective.

Toxic Atmospheres: When Air Becomes Weapon

Toxicity is atmospheric. Atmosphere is not just the measurable quality of air but also to the diffuse affective ambience through which environments are sensed and inhabited. Toxic air reconfigures space rendering breath precarious and produces sensorial cues such as sulphur, petrochemical haze, chemical dust, and the rot of decay, which signal danger

[1] The Bhopal disaster (1984) occurred when methyl isocyanate gas leaked from the Union Carbide pesticide plant in central India, killing thousands immediately and exposing over half a million people to toxic chemicals. Its aftermath continues through chronic illness, contamination of soil and groundwater, intergenerational health effects, and unresolved legal struggles.

[2] The Fukushima Daiichi nuclear disaster (2011) unfolded after a magnitude 9.0 earthquake and subsequent tsunami disabled cooling systems at the Tokyo Electric Power Company (TEPCO) plant, causing multiple reactor meltdowns. Radioactive materials were released into the atmosphere and the Pacific Ocean, generating a long-term environmental, political, and social crisis marked by displacement, contamination, and profound distrust in governmental and corporate management of risk.

[3] For example the Deepwater Horizon oil spill (2010, Gulf of Mexico), the Lac-Mégantic crude-oil train explosion (2013, Canada), the Samarco (2015) and Brumadinho (2019) tailings-dam collapses in Brazil, the Tianjin port explosion (2015, China), the Beirut ammonium-nitrate blast (2020, Lebanon), the East Palestine chemical train derailment (2023, USA), the Aliso Canyon methane leak (2015–16, USA), and contamination associated with European lignite mining in the 2020s. Each is widely analysed as a systemic corporate–state failure whose long toxic afterlives reveal that such disasters are not isolated events but manifestations of deregulated extraction, infrastructural decay, and environmental injustice.

even when contaminants remain invisible. Cultural texts consistently mobilise atmospheric toxicity to register landscapes and environments under siege. In Todd Haynes's *Safe* (1995), Julianne Moore's domestic world is slowly permeated by imperceptible pollutants, unravelling the fiction of suburban immunity. Bong Joon-ho's *The Host* (2006) opens with the dumping of chemicals into the Han River, atmosphericising harm across ecological and psychological registers. *Blade Runner 2049* (2017) frames its dystopia through toxic smog, irradiated ruins, and heavy yellow skies. In *The Last of Us* (2023–), fungal spores transform air into an unbreathable terrain, governing movement, intimacy, and risk. Works like *Dark Waters* (2019) and *Erin Brockovich* (2000) use atmospheric cues to index slow contamination before it becomes legible in law. Horror and gaming texts such as *Resident Evil* (2021) and *Silent Hill* (film 2006; games 1999–) harness toxic fogs, chemical mists, and airborne pathogens to evoke dread and instability, while *Mad Max: Fury Road* (2015) imagines a petro-apocalyptic future where dust storms, particulate sand, and chemical fogs make atmosphere itself the terrain of violence. Miyazaki's *Princess Mononoke* (1997) uses industrial atmospheres to point to the ecological violence of extractive modernity. Across these diverse media, atmosphere emerges as inseparable from feeling. Toxicity is apprehended through anxiety, suffocation, dread, and subtle forms of bodily unease that precede, exceed, or resist biomedical verification. Such affective responses underscore how harm is registered first somatically and phenomenologically, before it becomes legible within scientific or legal frameworks. At a planetary scale, Ulrich Beck's *risk society* (1992) reminds us that modernity generates globalised atmospheric hazards that collapse distinctions between local and global, present and future, cause and effect. Atmosphere names not only the surrounding air but the pervasive condition of inhabiting environments saturated by unseen risks of petromodernity in which toxicity becomes ordinary, ambient, and unevenly distributed across race, class, and geography. This planetary atmospherics reveals toxicity not as episodic crisis but as the ongoing background condition of contemporary life, shaping what can be breathed, sensed, and imagined.

Toxic Residue: Landscapes, Spaces, and Bodies That Remember

If atmosphere describes toxicity in its volatile, circulating forms, residue names a different ontology of toxicity. Residue is the material stubbornness of harm, the way toxicity congeals into ground, sediment, and infrastructure. Residue is what remains but also the harms that refuses to disappear. It is through residues that toxic events persist well beyond their supposed containment, demonstrating the non-linear temporality of industrial and colonial violence. *Contaminated diversity* (Tsing, 2012) captures the idea that residues do not immediately stop life, but reorganise it by producing damaged ecologies that are neither annihilated nor resilient, but caught in a suspended state of toxic flourishing. Residue thus functions as both material archive and political indictment. It records histories of extraction and state negligence, inscribing them in matter itself, where they circulate silently yet relentlessly across generations (Flyn, 2021; Nixon, 2011).

Globally, residue forms a cartography of harm. Canada's Athabasca tar sands constitute one of the most intensive and destructive sites of petro-capitalist extraction, transforming vast tracts of forest into an industrial residue-scape. Bitumen mining produces enormous tailings ponds, lakes of chemically complex waste containing heavy metals, naphthenic acids, and residual hydrocarbons that leach into groundwater and the Athabasca River, generating long-term toxic residues that accumulate in fish, sediments, and human bodies. These contaminants disproportionately affect downstream Indigenous communities of Fort Chipewyan, exemplifying the entwined dynamics of *slow violence* and colonial dispossession. The tar sands materialise the extended temporal span of extractive harm: even if extraction ceased today, the toxic burden would persist for centuries, embedding industrial history into land and water. Documentaries like *Petropolis* (2009) render these residues visually, portraying the tar sands as a vast infrastructural scar. The catastrophic nuclear accident that occurred on 26 April 1986 at Chernobyl's Reactor No. 4 resulted in an exclusion zone, a 30-kilometre radius around the plant where radioactive residues persist in soils, forests, water systems, fungi, and animal bodies, forming multispecies ecologies that challenge conventional ideas of contamination and survival. The Zone's abandoned towns, most notably Pripyat, serve as haunting infrastructures of residue. The abandoned mining town of

Wittenoom in Western Australia exemplifies the afterlife of extractive capitalism: asbestos fibres remain airborne decades after closure, circulating as lethal micro-residues that exceed any political attempt at erasure. Across Eastern Europe, derelict Soviet military bases leak toxins into groundwater, embedding Cold War geopolitics into the literal chemical memory of landscapes. These sites demonstrate that residue is not a by-product of modernity but it is one of its primary modes of existence.

The mundane temporalities of lingering residual exposure are also foregrounded in *Silkwood* (1983), which reframes contamination as bodily inscription: radioactive particles in Karen Silkwood's hair and home reveal toxicity not as an event but as an ongoing condition that reorganises subjectivity and domesticity. *True Detective* (Season 1, 2014) transforms Louisiana's Cancer Alley into a toxic archive, a place where petrochemical residue animates atmosphere, aesthetics, and fate. Tarkovsky's *Stalker* (1979) conceptualises residue metaphysically: the Zone is a terrain sedimented with past violence, where objects, ruins, and soils bear the imprint of events that cannot be fully known, only felt. Even franchise cinema such as *Godzilla Minus One* (2023) mobilises radioactive residue as a transgenerational wound, embedding nuclear testing into the affective and ecological memory of the nation. These landscapes exemplify *plutopias* (Brown, 2013) communities where nuclear and chemical residues pervade tissues, routines, and expectations to the point of becoming normalised. They expose the profound disjuncture between political time with its desire for closure, remediation, and narrative finality and material time, in which radioactive isotopes, heavy metals, and industrial chemicals persist for centuries or millennia. Residue thus exceeds the human lifespan.

When Disaster Becomes Domestic

Noah Baumbach's *White Noise* (2022) offers a contemporary examinations of the lived experience of toxic residue. Based on the 1985 novel by Don DeLilo which diagnosed the saturation of everyday life by media, catastrophe, and consumer desire, Baumbach's adaptation translates these anxieties into *mise-en-scène* where toxicity shapes atmosphere, embodiment, and domesticity in ways that exceed the human capacity to comprehend them. In doing so Baumbach adds is a visual grammar of the toxic residue of *petro-culture*. The world of the film is organised around everyday infrastructures of fossil-fuel modernity: the supermarket, the station wagon, the freeway, the university campus, and the chemical plant. These spaces are not incidental, they are the very architectures through which petro-modernity distributes risk and desire. Baumbach positions the supermarket as the aesthetic and ideological heart of this world, an environment of fluorescent plenitude that obscures the toxic residues enabling its abundance. The film's 'Airborne Toxic Event' is rendered as both spectacle and epistemological rupture. The billowing toxic cloud, and its paint-like quality of swirling purples, greens, and industrial blacks mirrors the *petro-sublime*: the awe and terror inspired by large-scale technogenic catastrophe. Yet Baumbach refuses to let the event remain a spectacular anomaly.

As particulate matter settles, the catastrophe transitions from atmosphere to residue. The disaster becomes ordinary. It seeps into groundwater, bodies, and routines. This movement from spectacle to sedimentation enacts *slow violence*, foregrounding toxicity as a temporally extended condition that resists narrative closure. Baumbach also underscores the epistemic instability inherent to toxic exposure. Information about the chemical cloud is contradictory, constantly revised, and mediated through news broadcasts, emergency bulletins, and rumour. The film's jittery oscillation between scientific jargon, bureaucratic confusion, and domestic panic captures the crisis of risk communication in late industrial societies demonstrating that when toxicity becomes too complex to narrate, it is displaced into fear, paranoia, and denial. Toxic embodiment is also made visible by Jack Gladney's exposure to Nyodene D, which becomes a metaphor for *petro-modern subjectivity*: to live in such a world is to be, to some degree, already contaminated. His body becomes a site of both chemical inscription and existential dread resulting in a recognition that mortality is no longer solely biological but ecologically mediated. The film turns toxic exposure into a form of ontological contamination, in which selfhood is inseparable from the infrastructures that sustain and imperil it.

Visually, the film leverages colour, saturation, and *mise-en-scène* to create a world where toxicity is both visible and invisible. The high-key lighting of the supermarket contrasts sharply with the atmospheric opacity of the toxic cloud, suggesting that modernity works by aesthetic partition: some forms of risk are hyper-visible, others are deliberately obscured. The incongruously beautiful cloud is emblematic of this dialectic renderering toxicity as visually seductive even as it poisons. *White Noise* stages a critique of petro-modernity that is not apocalyptic but banal, rhythmic, and cyclical. The film suggests that catastrophe has become routinised.

What is most unsettling is not the disaster itself but how seamlessly life resumes afterward in a return to consumption, spectacle, and everyday distraction. The toxic residue remains, but the cultural narrative moves on. *White Noise* is a diagnostic text for the present. It reveals how toxicity is lived: as atmosphere, residue, mood, anxiety, affect, and inheritance. It insists that petro-modern life is structured by exposures we barely perceive, and that the greatest violence of toxicity lies in how ordinary it becomes.

Toxic Time: When Exposure Becomes History

A further dimension of toxicity is time. Toxic time unfolds non-linearly, resisting chronologies in which causes and effects follow predictable sequences. It is marked by delays, lingering afterlives, and deferred recognitions that break the conventional coupling of exposure and illness, event and aftermath. In toxic temporalities, toxicity builds through repetition, leakage, or accumulation over time. Illnesses surface decades after exposure, residues persist long after industrial actors depart, and repair is endlessly postponed. Adriana Petryna's work on the aftermath of the 1986 Chernobyl disaster demonstrates that radiation produced not only biological damage but new social identities, as survivors were compelled to navigate bureaucratic, medical, and juridical systems in order to claim recognition as *biological citizens* (Petryna, 2004). Toxic exposure thus creates new temporal subjects, whose lives are structured by latency, uncertainty, and the long tail of harm. Communities living with contamination experience temporal dispossession (Nixon, 2011). Toxic suffering is obscured by political and economic systems calibrated to the short-term needs of profit and extractive gain. Indigenous communities in Australia's Northern Territory continue to fight for recognition of the long-term consequences of uranium mining, where radioactive and chemical residues persist in land and waterways long after companies withdraw. On Mirarr Country in Kakadu, the Ranger Uranium Mine (1981–21) recorded more than 200 contamination incidents, including the 2013 spill of 1.4 million litres of radioactive slurry. Independent monitoring has identified elevated concentrations of uranium, magnesium, and sulphate in Magela Creek and surrounding wetlands central to Mirarr cultural practices. Rehabilitation, mandated to be completed by 2026, is already projected to require decades of ongoing management, with scientists warning that hydrological flows continue to mobilise buried residues.

Further south, on Kungarakan and Warai Country, the legacy of the Rum Jungle uranium and copper mine (1954–71) demonstrates the extreme persistence of toxic time: acid mine drainage continues to leach uranium, copper, and sulphate into the East Finniss River more than 50 years after closure, despite multiple state-funded remediation efforts in the 1980s, 2000s, and the current federally backed rehabilitation plan (Mudd, 2008; Parliament of Australia, 2014). Across these sites, the temporalities of extraction far exceed the lifespan of industrial operations where residues remain biologically active, culturally damaging, and ecologically destabilising for generations, exemplifying the slow violence and temporal dispossession produced when extractive modernity imposes harms that unravel long after corporate and governmental actors retreat. Similarly, residents of Flint, Michigan, endure the protracted afterlife of lead-contaminated water, living with biological, infrastructural, and psychosocial decay long after the crisis was administratively 'closed' (Pulido, 2016). The Flint water crisis began in April 2014 when the city, under state-appointed emergency management, switched its municipal water source from Detroit's treated Lake Huron supply to the corrosive Flint River without implementing federally required corrosion-control treatment. The untreated water caused lead to leach from Flint's ageing pipe network, exposing an estimated 99,000 residents, predominantly Black, working-class communities, to dangerously high lead levels. Residents reported discoloured water, skin rashes, and hair loss, while independent testing revealed elevated blood-lead levels in children, findings initially dismissed by state officials. Even after the crisis was officially declared over in 2016, lead and bacterial contamination persisted, with pipe replacements ongoing for years. Flint residents continue to experience long-term neurological, developmental, and reproductive health impacts; economic decline; and profound mistrust in government. Flint is a paradigmatic case of environmental racism and the slow violence of infrastructural abandonment, where austerity governance imposed disproportionate toxic risk on already marginalised communities. Flint exemplifies how toxic time operates through denial and deferred responsibility, embedding racialised dispossession within the very infrastructures meant to sustain life.

Solarpunk and *biopunk* literatures offer critical imaginaries for understanding toxicity as a long-duration, multi-generational condition that shapes futures as much as presents. Solarpunk emerges as a speculative response to ecological breakdown, envisioning post-fossil fuel worlds oriented toward repair, renewable energy, and collective resilience, but always in dialogue with the residual damage of *petro-modernity* with its polluted atmospheres, degraded soils, and exhausted infrastructures that cannot simply be erased (Schuller, 2024). Its utopian impulses are grounded rather than escapist, built upon sedimented histories of harm and emphasising adaptation, care, and intergenerational responsibility rather than technological salvation. *Biopunk,* by contrast, approaches toxicity through the body itself (Schönbohm & Hermann, 2024). Drawing on traditions of *cyberpunk* and biopolitical critique, biopunk narratives imagine futures shaped by genetic modification, contamination, pharmaceutical regimes, and corporate control over life processes, where bodies become sites of toxic inscription and experimental governance (Braidotti, 2013; Vint, 2007). In these worlds, exposure is not accidental but systemic, revealing how environmental toxicity, biological vulnerability, and political power converge. Together, solarpunk and biopunk foreground toxicity not as an episodic disaster but as an enduring condition that reorganises landscapes, bodies, and temporal horizons, demanding new ethical and narrative frameworks for living with inherited harm.

Works by Paolo Bacigalupi, N. K. Jemisin, and Octavia Butler epitomise these toxic temporal aesthetics. In Bacigalupi's *The Windup Girl* (2009), environmental collapse and gene-hacked agricultural systems create a world where toxic residues determine who survives. Modified crops, engineered plagues, and corporate monopolies on clean seed render generational survival as an ecological and political lottery. Similarly, Jemisin's work, especially *The Broken Earth* trilogy, renders toxicity geological and intergenerational. Here, the Earth itself is a wounded, volatile body, its toxic storms and seismic upheavals functioning as the long afterlife of extractive histories and anthropogenic violence. Contamination is transmitted across millennia, structuring kinship, labour, and the distribution of power. Octavia Butler's *Parable* series (*Parable of the Sower*, 1993; *Parable of the Talents*, 1998) situates toxicity within the socio-political collapse of climate catastrophe, environmental racism, and corporate enclosure. Harm unfolds slowly as chemical residues accumulate, fertility declines, and communities fracture under the weight of ecological precarity. Butler's generational protagonists navigate a world where toxic time is not apocalypse but ongoing condition where survival demands new cultural, spiritual, and ecological practices attuned to damaged environments. Both *solarpunk* and *biopunk* articulate a future shaped by toxic residue, colonial extraction, and the uneven temporalities of harm.

Science fiction, in particular, excels at imagining toxic temporalities that exceed human time, projecting ecological injury, corporate extraction, and biological contamination into futures measured not in decades but in evolutionary and planetary scales. *Alien: Earth* (2024–25) amplifies this logic by bringing the franchise's long-standing concerns with biocorporate violence back to the home planet, transforming Earth into a site where contamination is no longer an externalised threat from the frontier but an internalised condition of everyday life. The film frames toxicity as a multi-generational inheritance, showing how the residues of corporate experimentation, bioweapon leaks, and airborne spores go on altering ecosystems long after the corporation's activities have been obscured or erased from institutional memory. The film also foregrounds how biotechnological pollution survives political regimes, economic cycles, and human lifespans. This aligns the franchise with environmental humanities scholarship by showing how petrochemical, genetic, and military residues extend far beyond their original moment of release. *Alien: Earth* thus is a meditation on the Anthropocene as a residue condition, where corporate science and planetary degradation converge to create environments inhospitable to future generations. Toxic time is therefore also cosmic, extending beyond human histories into evolutionary and post-human futures.

Films such as *Arrival* (2016) imagine non-linear temporalities where contamination be it linguistic, biological, or spatial transforms perception and memory. *Children of Men* (2006) frames infertility as a slow global toxicity, a reproductive residue of political and environmental decay. *Dark* (2017–20) literalises temporal contamination, depicting time as a looping ecology in which past contamination returns to structure the future. Across these works, toxic time emerges not as supplementary to toxic landscapes but as their organising principle. Toxicity is anticipatory (a future harm already shaping the present), belated (a past injury returning as illness decades later), and intergenerational (a residue transmitted through DNA, groundwater, or political neglect). Toxic time forces a rethinking of accountability, governance, and care. It compels recognition that the harm of petro-modernity is not bounded by the event but unfolds across long arcs of duration, embedding violence in the very temporal fabric of contemporary life.

Cinematic Toxic Atmospheres

To examine contaminated ground through film and television is to recognise how screen media provide some of the most incisive ways of perceiving toxicity. Film and television have a long history of creatively rendering polluted atmospheres, poisoned soils, and toxic futures through composition, lighting, colour, sound, and *mise-en-scène*. Toxic landscapes on screen are not just representational, they are formal ecologies, shaped through aesthetic decisions that make harm legible at the level of sensation, spectacle of sublime. As such cinematic toxicity is a world-making aesthetic, a cinematic infrastructure that shapes how viewers are taught to see, hear, and feel contaminated ground. Screened toxicity is never neutral. Its visual grammar carries political weight, encoding histories of extraction, colonial abandonment, and state-sanctioned sacrifice zones. By attending to cinematography, sound design, framing, colour palettes, and spatial composition, film and television construct toxic landscapes as experiential worlds, worlds where contamination is both narrative content and aesthetic structure.

Atmosphere is one of the most powerful cinematic tools for rendering toxicity. Whether captured as particulate haze, petrochemical fog or radioactive air, atmospheric contamination often appears first not in plot but in colour temperature, diffusion, visual noise, and depth-of-field distortion. In screen culture, toxic atmospheres are sensed through form. In Jia Zhangke's *Still Life* (2006), the toxic atmosphere seeps directly from the socio-ecological upheaval of the Three Gorges Dam project,[4] where mass displacement, demolition, and inundation have transformed entire towns into transitional ruin. The region's air is rendered as a perpetual haze, a drifting mixture of dust from collapsing buildings, humidity rising off the newly formed reservoir, and the industrial emissions of accelerated redevelopment. Cinematographer Yu Lik-wai deploys

[4] The Three Gorges Dam, constructed on the Yangtze River between 1994 and 2012, is the world's largest hydroelectric project and a defining symbol of China's state-led modernisation. Its creation flooded more than 600 square kilometres of land and displaced an estimated 1.3–1.5 million residents, producing long-term social upheaval and economic precarity. Ecologically, the dam has altered sediment flows, intensified landslides and seismic risks, reduced downstream biodiversity, and contributed to the extinction or endangerment of several Yangtze species. Submerged industrial sites and waste dumps have released pollutants into the reservoir, while fluctuating water levels and ongoing construction generate persistent atmospheric and particulate contamination.

muted greens, washed-out blues, and overexposed whites, alongside a thin digital grain that unsettles the image's clarity. This aesthetic registers toxicity as a sensorial weight, a heaviness that presses onto bodies, landscapes, and the film's own visual texture. The dissolving boundary between sky and concrete underscores a world where demolition dust becomes weather. Toxicity here is not visual spectacle but ambient suffocation, a quietly overwhelming condition that conveys the disorientation, loss, and corporeal strain produced by China's hydropower modernity and its relentless remaking of ecological and social time.

Mohammad Rasoulof's[5] *There Is No Evil* (2020) is structured as a quartet of interconnected but distinct stories providing a mosaic portrait of how state violence infiltrates everyday life. The film uses domesticity, rural isolation, and intimate relationships to show that the machinery of state violence (specifically execution) permeates homes, landscapes, and the moral fabric of the nation. The film's atmospheric palette of dust-laden yellows, bleached greys, and smog-thickened skies echoes the pervasive environmental contamination characteristic of Iran's *petro-state economy*. Rasoulof's use of diffuse light and hazy horizons performs a dual function: formally, it flattens depth of field and gives the air a grainy density that suggests particulate pollution; thematically, it renders the landscape itself complicit in the film's critique of authoritarian governance. The visual suffocation of the atmosphere mirrors the suffocation of dissent, linking ecological degradation to state repression. Industrial backdrops, desiccated plains, and polluted cityscapes become moral topographies, where environmental toxicity is inseparable from political toxicity. In this way, Rasoulof extends the *environmental humanities'* insight that atmosphere is not merely meteorological but ideological and that the air of the nation becomes a medium through which coercion, fear, and constraint are breathed and lived.

A further illustration of atmospheric toxicity as world-building appears in Anocha Suwichakornpong's Thai *By the Time It Gets Dark* (2016), a film that intertwines political memory with ecological degradation. Although the plot moves elliptically across characters, historical moments,

[5] Mohammad Rasoulof was born in Shiraz, Iran, and is an internationally recognised Iranian filmmaker whose work has frequently critiqued state power, censorship, and social repression. His films have led to repeated bans, travel restrictions, and legal persecution by Iranian authorities, making his nationality and political context central to readings of his work.

and shifting realities, the film consistently uses atmosphere to register the lingering contamination of Thailand's political and environmental landscape. Much of the film unfolds in humid, pesticide-treated agricultural zones, where a pervasive yellow-green haze flattens spatial depth and produces a visual sense of damp suffocation. Cinematographer Ming Kai Leung bathes scenes in diffused, low-contrast light, creating images that seem to sweat, blur, and shimmer under the pressure of heat and particulate density. This atmospheric heaviness operates as a sensory analogue to the film's political subject: the unprocessed trauma of the 1976 Thammasat University massacre.[6] In Suwichakornpong's world, air is the carrier of both chemical residues from industrial farming and spectral residues of political violence. The paraffin-lit interiors, smoky rural vistas, and polluted city edges work together to generate a cinematic ecology in which toxicity is experienced as disorientation, stagnation, and repetition, mirroring the stagnation of political reform in Thailand. Here, atmosphere is not an environmental backdrop but an affective climate that conveys how historical violence persists as a breathable, enveloping condition, a toxic inheritance that saturates both landscape and memory.

African filmmakers deploy atmosphere to reveal postcolonial toxicity with particular acuity. In Abderrahmane Sissako's *Bamako* (2006), the entire narrative unfolds within the dusty courtyard of a residential compound in the titular city located in Mali, where an improvised tribunal prosecutes the World Bank and IMF for the structural violence of neoliberal policy in Africa. This modest, everyday urban location frames cracked earth, sun-bleached walls, drifting sand, and encroaching dryness as a cinematic microcosm of contaminated ground. The *Harmattan*[7] haze,

[6] The 1976 Thammasat University massacre occurred on 6 October 1976 in Bangkok, when Thai state security forces, police units, and right-wing paramilitaries attacked student protesters occupying Thammasat University in opposition to the return of former dictator Thanom Kittikachorn. The assault involved live ammunition, lynchings, and the torture of students; official counts listed 46 dead, though independent estimates suggest far higher casualties. Thousands were arrested. The violence marked a decisive backlash against Thailand's student democracy movement and ushered in a period of military rule and political repression. Long suppressed in official discourse, the massacre endures in Thai cultural memory as a traumatic episode of state brutality, censorship, and anti-communist hysteria.

[7] Harmattan refers to a dry, dust-laden trade wind that blows from the Sahara Desert across West Africa between November and March. Characterised by low humidity, reduced visibility, and large quantities of fine particulate dust suspended in the air, Harmattan often blankets cities and rural regions alike in a yellow-grey haze. The wind can lower temperatures at night while intensifying heat and aridity during the day, contributing

captured as a persistent yellow-grey veil functions as more than seasonal weather, it is framed as a materialisation of economic and political toxicity, a particulate atmosphere laden with the aftereffects of debt, austerity, and imposed development schemes. Sissako's slow pans across the landscape reveal domestic life unfolding with women dyeing fabric, men repairing saddles, and children weaving through the proceedings suggesting that the contamination of everyday life by global finance is as insidious and pervasive as airborne dust. The film's atmospheric opacity encodes the residues of extractive globalisation, turning the landscape into a world where the air itself bears witness to decades of privatisation, resource depletion, and state erosion. In this sense, *Bamako* renders toxicity both literally and metaphorically with the dusty ground is a lived archive of postcolonial harm, while the suffocating air becomes an index of neoliberal contamination that infiltrates lungs, landscapes, and political futures. In South Africa's *Moffie* (2019), director Oliver Hermanus uses desert atmospheres of bleached, overexposed light, swirling dry dust, and the chemical haze of weapons training fields to render apartheid's militarisation as an environmental condition. The film's visual palette evokes an air that is as abrasive and hostile as the army's disciplinary regime. Drill routines unfold in heat-shimmered landscapes where dust clings to skin and uniforms, suggesting that violence is inhaled as much as enacted. The harshness of the land mirrors state brutality while the toxic atmosphere registers how apartheid's racism, homophobia, and authoritarianism seep into the very air the soldiers breathe. In this way, Hermanus frames environment not as backdrop but as a vector of psychic and political contamination.

Cinematic Toxic Residue

If atmosphere gives toxicity its cinematic air, residue anchors contamination in the materiality of ground, in dirt, mud, metal, concrete, bone, and chemical sediment. Films across the world demonstrate how toxic residues operate as geological inscriptions of political and economic violence. Latin American cinema frequently situates toxicity in land dispossession and

to respiratory illnesses, crop stress, and infrastructural wear. In cultural and cinematic contexts, Harmattan frequently appears as a sensory marker of environmental harshness, atmospheric opacity, and ecological vulnerability, conditions that can be mobilised metaphorically to signify broader forms of social and political toxicity.

extractive economies. Pablo Larraín's *Neruda* (2016) uses a landscape of nitrate mines and abandoned encampments to visualise the toxic legacies of Chile's extractive industries. Northern Chile's Atacama Desert is shot as a pale, irradiated wasteland whose residues recall Patricio Guzmán's documentaries about Chile's nitrate ghost towns. Guzmán in *Nostalgia for the Light* (2010) juxtaposes astronomers studying the cosmos with archaeologists and families searching for the remains of the disappeared in the desert's nitrate fields. The Atacama's nitrate ghost towns, once booming company settlements during the late nineteenth- and early twentieth-century saltpetre boom[8] now stand as barren ruins, their corroded structures and mineral-encrusted soils recording the violence of resource extraction, labour exploitation, and the later use of the desert as a site of torture and clandestine graves during the Pinochet dictatorship. In both representations, the nitrate ghost towns function as contaminated memoryscapes: geologies that preserve the residues of capitalism, militarism, and environmental degradation. Their abandoned infrastructures and dusty atmospheres embody the idea that landscapes can retain and transmit historical trauma long after the extractive economies that created them have vanished.

In *La Tierra y la Sombra* (2015), César Acevedo situates a rural family within a Columbian landscape overwhelmed by industrial agriculture, where the burning of sugarcane fields produces a constant fall of ash that coats bodies, crops, interiors, and air alike. The film's static, oppressive composition, and monochrome greys emphasise how particulate matter becomes an everyday hazard, transforming labour into a form of toxic exposure that seeps into domestic space. A related atmospheric toxicity underlies Fernando Eimbcke's *Lake Tahoe* (2008), whose minimalist narrative belies a Mexican landscape quietly shaped by pollution, automotive debris, and contaminated shoreline. Eimbcke's slow pans across industrial peripheries and chemically dulled surfaces frame adolescence against a geography permeated by infrastructural waste. These films reveal how Latin American cinema frequently registers contamination not

[8] Saltpetre (also spelled *saltpeter*) is a common name for sodium nitrate or potassium nitrate, chemical compounds historically used in fertilisers, explosives, glassmaking, and food preservation. In the late nineteenth and early twentieth centuries, northern Chile became the global centre of saltpetre (nitrate) mining, with the Atacama Desert supplying the world's most important natural source of nitrates before the advent of synthetic fertilisers.

through spectacle but through ambient, insidious forms such as ash, dust, and residue that infiltrate the spaces of home, family, and youth, rendering toxicity a lived and intimate condition rather than a distant environmental crisis.

Cinematic toxic residue appears in Todd Haynes's *Safe* (1995). Set in suburban Los Angeles, it renders the domestic environment as a contaminated ecology. The film's immaculate lawns, manicured gardens, and pastel interiors are quietly saturated with pesticides, cleaning chemicals, and airborne irritants, everyday residues of late-twentieth-century American consumer culture. Haynes doesn't depict pollution through spectacle but through a slow accumulation of invisible threats: vapours from dry-cleaned clothes, fumes from car exhaust, dust from construction sites, and household aerosols that linger in the air and settle onto furniture, clothing, and skin. The protagonist Carol develops increasingly debilitating sensitivities that the medical establishment refuses to recognise, highlighting how toxic residues in suburban spaces are medically negated rather than identified as environmental harm. The film's shallow-focus shots of sterile kitchens, beige carpets, and chemically maintained gardens emphasise a world where residue is aestheticized, disguised by the clean surfaces and controlled environments of middle-class domesticity. *Safe* thus reframes American suburbia as a toxic ground with its apparent order built upon layers of chemical sedimentation that infiltrate bodies and reshape subjectivity. In this way, the film extends the concept of toxic residue beyond industrial ruins and extractive sites, revealing the intimate and insidious forms of contamination that settle quietly into everyday life.

Cinematic Toxic Time

Cinema is uniquely capable of rendering toxic temporality and its soundscapes through editing rhythms, narrative ellipses, temporal loops, and long takes. Toxic time operates in screen cultures as an aesthetic analogue to environmental latency. Péter Mészáros's *The Gaze of the Gorgon* (2013) exemplifies toxic temporality through its meditative tracking shots across abandoned Hungarian factories and the desolate peripheries of the post-Soviet industrial landscape. Rusted cranes, collapsed cooling towers, and irradiated concrete slabs appear in extended, near-static compositions that force viewers to dwell in duration rather than dramatic action. This slowness functions as formal critique: the long take becomes an ethical demand to attend to the residues of industrial glory that now persist as ruin. The film's temporal juxtaposition of past modernist ambitions sitting with present dereliction exposes the afterlife of extraction as a condition in which time itself becomes polluted, dragging forward the consequences of a vanished industrial epoch.

A similar rendering of toxic latency structures Valentin Hotea's *Roxanne* (2013). Although the film concerns political surveillance rather than environmental collapse, it frames Romanian state files as toxic archives that belatedly infiltrate the lives of the living. Secret policing, betrayal, state scrutiny reemerge decades later, reorganising relationships, personal identities, and national memory. The film's temporal structure mirrors environmental contamination: harm is stored, dormant, and then released without warning. In Hotea's hands, toxic time becomes a metaphor for authoritarian residue, demonstrating how the machinery of surveillance, like chemical pollutants, can persist invisibly until a moment of sudden and destabilising return. Ása Hjörleifsdóttir's *The Swan* (2017) offers yet another articulation of toxic time through a coming-of-age story set amid geothermal vents, sulphuric landscapes, and the creeping presence of industrial steam. Although the narrative centres on childhood, the surrounding Icelandic environment introduces a seasonally shifting toxic temporality. Whether through the Soviet-era ruins of Hungary, the archival hauntings of post-communist Romania, or the geothermal-industrial landscapes of Iceland, each film demonstrates how contamination folds the past into the present and troubles the future. Toxicity becomes a temporal force rendered visible through cinema.

Toxic time in Debra Granik's *Winter's Bone* (2010) is structured differently where the rural Ozarks are shaped by the slow, generational fallout of methamphetamine production rather than a singular toxic event. The film's barren woods, collapsing houses, rusting cars, and chemically scorched clearings register a landscape marked by accretive harm, not explosive catastrophe but the long-term seepage of chemical processes associated with illegal drug manufacture. Granik's use of grey winter light, extended silences, and lingering shots of ruined outbuildings and contaminated yards reinforces a temporality of attrition: time in the Ozarks does not advance so much as erode. People's bodies, like their houses, show the effects of prolonged exposure to poverty, to toxins, and to the corrosive demands of survival. Crucially, the film's narrative unfolds at the pace of toxic latency: past actions including abandoned meth labs and familial debts resurface belatedly to jeopardise the present. Secrets buried in the soil of community re-emerge like radioactive isotopes, unpredictable and destabilising. In *Winter's Bone*, toxic time becomes a form of rural haunting, where each season of snowmelt reveals new residues and the future remains constrained by the chemical, economic, and emotional contaminants inherited from the past.

Cinematic Toxic Soundscapes

Sound is not just cinematic atmospherics. Sounds are epistemologies, embodied ways communities register slow harm and make it legible. In cinematic representation, sound often reveals what sight cannot: in contaminated places, sonic cues become diagnostic. Rachel Carson's *Silent Spring* (1962) remains the foundational narrative of sensory harm: toxicity announces itself not through spectacle but through absence; the bird that no longer sings, the spring that no longer resounds. This negative acoustics has been widely adopted in cinema to temporalise exposure, rendering harm not as an explosive event but as a durational silence that accumulates meaning over time. Toxicity in screen cultures is often heard before it is seen. Sound designers use hums, drones, metallic resonance, and industrial rumbles to evoke contamination as an auditory condition. Yet sonic design is one of the most effective but often overlooked ways that film renders toxicity. In *Beijing Bicycle* (2001), the incessant hum of traffic, grinding of machinery, and clatter of urban construction crafts a toxic sonic ecology that mirrors China's accelerated industrialisation. The city's acoustic density of honking horns, metallic brakes, and perpetual background drone becomes an auditory residue of infrastructural expansion. Viewers do not simply see pollution, they hear it pressing against bodies and space, mapping industrial modernity as a condition that vibrates through the air.

Julius Onah's *The Cloverfield Paradox* (2018) mobilises deep rumbles, metallic groans, and unsettling resonances to imagine a world where technological experimentation destabilises space-time. Although science fiction, the film's soundscape borrows from the auditory textures of collapsing industrial infrastructures including distant booms, structural buckling, and pressure alarms creating a sonic atmosphere that evokes contamination and systemic failure. This alignment of technological hubris with acoustic disturbance situates toxicity not only in narrative content but in the vibrational quality of the film's world. Animation pushes the trope further. In Pixar's *WALL·E* (2008), Earth is rendered as a planetary landfill where mechanical whirs, robotic bleeps, and the clatter of debris replace birdsong. The film's sound design literalises the central insight of *ecocinema studies* that popular forms can train audiences to perceive environmental degradation as sensory transformation, even if the affective packaging risks domesticating critique.

Television also mobilises a grammar of absence via sound. HBO's *The Last of Us* (2023) situates its fungal apocalypse in hushed, spore-laden environments where silence itself signals danger. The near-total absence of human bustle is broken only by distant creaks, muffled coughs, or the clicking of infected bodies, producing an atmosphere in which the world's contamination is heard as much through what has vanished as through what remains. Like Carson's vanished birdsong, these silences mark a poisoned ecology where sound, or its loss, becomes the most immediate evidence of planetary unliveability. A quieter but equally potent mode of sonic toxicity appears in Abbas Kiarostami's Iranian film *The Wind Will Carry Us* (1999), where the conspicuous absence of birdsong, insects, and other nonhuman sounds generates a sonic void that gradually acquires an unsettling ecological significance. The stillness of the rural village is not the pastoral quiet of an unspoiled environment, rather it is charged with the sense that something is missing. Kiarostami's minimalistic soundscape broken only by the hum of distant cars, the scrape of a spade, the faint ring of a mobile phone renders silence a form of acoustic residue, the auditory trace of an ecosystem under strain. By withholding expected environmental cues, the film reconfigures silence as a toxic signal, aligning the absence of nonhuman sound with its thematic interests in mortality, impermanence, and the precariousness of rural life. In Kiarostami's hands, toxic time is heard before it is seen. The landscape's quiet registers a temporal thinning suggesting that ecological decline is already underway, even if it escapes explicit representation.

A similar sonic toxicity appears in Camille Henrot's short film *Grosse Fatigue* (2013), which uses archival animal recordings and their gradual disappearance to posit ecological exhaustion. The film layers snippets of birdsong, insect buzzing, and mammalian calls drawn from natural history archives, only to interrupt, distort, or abruptly cut them as the visual montage accelerates. These truncated animal sounds create a sense of sonic extinction in progress, as if the species themselves are being silenced within the filmic frame. The resulting auditory instability mimics the collapse of multispecies worlds under the pressure of scientific extraction, anthropological cataloguing, and global biodiversity loss. This fading of animal sound functions as a toxic temporal marker acting as a reminder that species decline is experienced first acoustically through thinning choruses and interrupted frequencies well before visual disappearance becomes evident. Henrot's film captures the ecological insight that sound is often the earliest index of crisis.

In John Krasinski's *A Quiet Place* (2018), animal sound is used to dramatise a world in ecological decline. Early scenes reveal a landscape in which birdsong has vanished entirely. The silence of the forest and riverbanks is evidence of a collapsed ecosystem. The few remaining animal calls from frogs and insects are so faint and sporadic that they become markers of vulnerability rather than vitality. In this way, *A Quiet Place* constructs a soundscape where the eradication of animal sound becomes a material sign of toxic futurity: the land is not only dangerous but emptied of its multispecies life, making silence itself an index of ecological devastation. Sonic toxicity also emerges through temporal lag in Apichatpong Weerasethakul's *Tropical Malady* (2004). As the narrative shifts from a queer romance in a Thai city to a mythic confrontation in the jungle, the soundscape thins dramatically: insects and birds fade, nocturnal calls become intermittent, and long stretches of the jungle fall into uncanny quiet. The absent soundscape marks a temporal breach, the collapse of linear time into the cyclical rhythms of animism and haunting. Here, silence becomes a portal into toxic temporality, where the forest seems to hold its breath in anticipation of an event that never quite arrives.

A different sonic strategy is employed in Lucrecia Martel's *La Ciénaga* (2001). Set in a decaying provincial Argentinian household during a heatwave, the film's sound design layers broken electrical fans, clattering shutters, distant thunder, and the incessant drone of insects. These sounds create an auditory humidity, evoking stagnation and rot. But crucially, the soundtrack contains gaps, moments where the cicadas drop out, the storm stalls, and the air goes still that function like sonic fissures. These pockets of silence signal the slow collapse of the natural environment under the pressure of class decay, alcoholism, and familial dysfunction. Toxicity emerges as both atmospheric and auditory: a rumbling undercurrent that never erupts into catastrophe but slowly suffocates the space. A more overtly ecological example is found in Ciro Guerra's *Embrace of the Serpent* (Columbia, 2015), where the Amazonian soundscape, rich with birds, insects, and river currents slowly loses its vitality as colonial and scientific incursions intensify. Guerra strategically modulates the jungle's sonic density. Early sequences pulse with multispecies vibrancy, while later scenes, set near rubber extraction camps and missionary outposts, become eerily quiet. The fading of nonhuman sound becomes an index of colonial extraction itself. Silence is not absence but record, a trace of multispecies loss, mirroring how environmental degradation registers first through subtle acoustic diminutions before visual signs appear.

A comparable logic underpins the dystopian soundscape of Alfonso Cuarón's *Children of Men* (2006). While not conventionally ecological, its sonic design mobilises urban silence as a marker of species crisis. The near-total absence of children's voices in playgrounds, schoolyards, and city streets becomes a form of anthropogenic sonic extinction. The empty sonic register is as devastating as the film's visual ruins: it anticipates a world in which the very frequencies of human social life have been depleted. Here, the toxic residue is temporal and generational, heard in the missing sounds of the future. Another example of weaponised silence appears in Hirokazu Kore-eda's *After Life* (Japan, 1998), which uses acoustic minimalism to represent worlds caught between life and afterlife. The quiet hallways, muted natural sounds, and occasional mechanical hums create an atmosphere of temporal suspension. This silence carries a mournful toxicity: the residues of lived experience accumulate like dust in the film's stillness, producing an ambient melancholy that speaks to the fragility of memory and the thin line that separates existence from erasure.

Cinematic Toxic Senses

Cinema and television repeatedly mobilise odour as a register of environmental harm and social abandonment. In *Gasland* (2010), smell becomes the earliest and most persistent indicator of toxic exposure. Residents living near hydraulic fracturing sites describe overwhelming rotten-egg stenches (the sulphuric odours associated with methane leaks and hydrogen sulphide emissions) permeating homes, wells, and surrounding landscapes. Others speak of a sickly chemical sweetness in the air, an atmosphere so noxious that it induces nausea, headaches, and forces families to abandon their houses altogether. These testimonies position smell as an early-warning system: a sensory mode of environmental knowing that precedes, and often contradicts, scientific measurement, regulatory thresholds, or official reassurance. Fox's film repeatedly foregrounds this olfactory knowledge through close, unmediated encounters: residents gag on fumes, cover their noses, and describe living in a state of constant sensory assault. Odour here is not incidental but diagnostic. It signals contamination before it can be stabilised as data, translating invisible chemical processes into embodied certainty. This sensory logic culminates in the film's most iconic moment, the ignition of tap water, where olfactory suspicion is confirmed visually and materially. The flaming faucet condenses smell, sight, and touch into a single event, revealing that what smells wrong is, indeed, materially unsafe. By foregrounding odour, *Gasland* demonstrates how toxic residue travels not only through groundwater and air but through sensory experience, affect, and cultural memory.

Television develops this sensory register with equal force. HBO's *Chernobyl* (2019) repeatedly invokes metallic taste and acrid odours reported by workers and first responders, sensory markers of radiation poisoning that official science and Soviet state secrecy sought to downplay. These evocations of odour turn smell into evidence of contamination, reminding audiences that the body's sensorium can outpace bureaucratic instruments. In a very different register, *The Simpsons* (1989–) has for decades used olfactory jokes to satirise environmental sacrifice zones: Springfield is constantly engulfed by the stench of its nuclear power plant, landfills, or industrial runoff. Mr. Burns boasts of "the sweet smell of progress" as sulphurous smoke billows, while townspeople gag at the air. Here, odour functions as a comic shorthand for disposability, but it also embeds the reality of toxic atmospheres into popular consciousness. The humour normalises contamination while simultaneously highlighting its absurdity, teaching audiences to recognise smell as the signature of corporate negligence and political indifference. Jordan Peele's *Us* follows a family on a holiday at Santa Cruz, which is disrupted when a group of doppelgängers, known as the Tethered, emerge from subterranean tunnels to terrorise them. As the Tethered rise across the country to kill and replace their surface counterparts, the film reveals that they are the abandoned products of a failed government experiment. The film deploys subterranean stench allegorically: the tunnels where the Tethered live are saturated with odour, mildew, and rot, constructing an olfactory dramaturgy of disposability in which marginalised populations are literally consigned to the realm of breathable waste.

Smell here becomes a social marker, an index of abjection that separates those who live above ground from those forced to inhabit toxic underworlds. Similarly, in John Hillcoat's *The Road* (USA, 2009), characters' visceral reactions to corpses, ash, and smoke stage smellscapes: olfactory cues that biopolitically organise exposure and vulnerability. In *The Killing Fields* (1984), odour becomes a visceral index of historical trauma. As journalist Sydney Schanberg and his Cambodian colleague Dith Pran move through the aftermath of Khmer Rouge atrocities,[9] smell is displaced onto actors' bodily responses as they recoil from the stench of mass graves, decomposing bodies, stagnant water, and scorched earth. The film repeatedly uses characters' expressions of nausea, gagging, and avoidance to convey a sensory landscape saturated by death. Smell becomes a mnemonic force, marking the landscape as a contaminated archive. The olfactory dimension underscores the politics of remembrance: the land is not only stained but suffused with the unburied residues of genocide. This use of smell extends how odour communicates not industrial toxicity alone but the toxic afterlives of political violence.

Offering a different register of toxic landscapes is Rolf de Heer and Peter Djigirr's *Ten Canoes* (2006) offers a strikingly different register of olfactory toxicity by drawing on Yolŋu relationships to Country, where smell operates as an intimate barometer of ecological and social balance. The film contrasts the odours of healthy Country: the earthy sweetness of eucalyptus smoke, the damp richness of wetlands, the vegetal scent of bark and paperbark shelters with the implied smells of death, sickness, and colonial intrusion, which linger at the edges of the narrative even when not directly shown. Yolŋu storytelling practices emphasise smell as a mode of orientation and knowledge-making, and the film's immersive soundscape and visual rhythms invite viewers to imagine these olfactory cues as part of a living ecology. When conflict, transgression, or impending harm enters the story, the environmental sensorium subtly shifts. The air seems heavier, the wetlands quieter, the smoke less cleansing, signalling a disturbance in Country's equilibrium. In this context, smell becomes a culturally specific indicator of whether land is cared for, respected, and in good relation, or whether it is under threat from internal disharmony or the slow encroachment of settler colonialism. *Ten Canoes* thus expands the framework of cinematic toxic senses by showing how, in Indigenous

[9] The Khmer Rouge atrocities refer to the systematic violence committed by the communist regime led by Pol Pot, which ruled Cambodia from 1975 to 1979. During this period, an estimated 1.7 to 2 million people, nearly a quarter of the country's population, died from execution, forced labour, starvation, and disease as the regime attempted to create an agrarian, classless society. Intellectuals, ethnic minorities, religious leaders, and perceived political opponents were targeted. The atrocities left enduring physical and psychological scars on the population and landscape, shaping Cambodia's post-conflict reconstruction and collective memory.

epistemologies, olfaction is not merely biological but relational, a way of sensing when Country itself is healthy or suffering.

Cinematic Tastelessness

Tastelessness emerges in cinema and television as a sensory index of contaminated ground, signalling the depletion of soils, ecologies, and food systems under toxic modernity. When landscapes are exhausted or chemically altered, flavour itself becomes a casualty, revealing environmental harm at the most intimate scale: the mouth. While smell and sound are key sensory registers through which cinema and television render toxicity legible, taste, or more precisely, the erosion of taste offers a powerful yet underexamined mode for representing toxic landscapes. Across global screen cultures, tasteless food appears not merely as a narrative detail but as a material sign of depleted ecologies, contaminated soils, industrial overproduction, and the disintegration of multispecies life. Taste, which depends upon healthy soil, clean water and biodiversity, becomes in cinema a fragile sensory threshold through which crisis is felt in the most intimate way via the mouth, on the tongue, and within the body.

One of cinema's earliest environmental dystopias, Richard Fleischer's *Soylent Green* (1973), uses tastelessness to dramatise the catastrophic consequences of climate change, overpopulation, and soil exhaustion. The mass-produced wafers that feed the population are described as flavourless, nutritionally minimal, and texturally dead. Their sameness underscores a world in which biodiversity has collapsed and agriculture has become impossible. When protagonist Sol Roth tastes fresh strawberries, beef, and vegetables, rare black-market luxuries, he weeps with recognition. Flavour becomes a lost archive of the world's former ecological abundance, a sensory memory of environments that no longer exist. The scene positions taste as a form of environmental mourning. In this context, the absence of taste materialises the slow violence of environmental degradation: flavour becomes scarce because the earth has been stripped of its capacity to nourish anything, and tastelessness becomes evidence of a planet whose soil has been rendered mute.

Tastelessness in film often functions as a sensory index of ecological collapse. In Pixar's *WALL·E* (2008) humanity survives on colour-coded, liquified meal replacements, nutrient slurries engineered for efficiency rather than pleasure. In the sterile interiors of the Axiom spaceship, food has been stripped of flavour, texture, and origin. The absence of taste allegorises a world in which the Earth itself has become a toxic landfill, its soils unable to sustain agriculture. The loss of flavour is not incidental; it is a signal that the ecological web that produces taste (soil, water, pollinators, biodiversity) has vanished. Tastelessness thus becomes the sensory residue of environmental ruin, a condition produced by the failure of ecological systems to support life. A similar erasure of taste structures Phillip Noyce's *The Giver* (2014), adapted from Lois Lowry's novel, where meals are deliberately bland, uniform, and nutritionally engineered. In this sanitised society, food's neutral flavour reflects the broader suppression of memory, difference, and ecological relation. Tasteless food embodies environmental amnesia. The class politics of toxic taste are foregrounded in Bong Joon-ho's *Snowpiercer* (2013). In the train's tail section, survivors consume protein blocks which are gelatinous, indistinguishable, and aggressively bland. Their tastelessness is part of a broader system of deprivation, a sensory marker of classed environmental inequality. Food is not merely flavourless it is also anonymous, its composition concealed until its horrifying origin is revealed. Taste in *Snowpiercer* is a social indicator. Those with access to flavour inhabit the front of the train, while the tasteless rations of the poor mark their disposable status. Blandness becomes a political sign of toxic futurity, a condition in which environmental collapse and authoritarian hierarchy converge. The critique of industrialised taste extends to Bong Joon-ho's *Okja* (2017), where genetically engineered super-pigs produce a hyper-efficient meat product designed to be flavour-neutral so that it can be universally marketable. Taste is sacrificed for profit, and flavourlessness becomes a symptom of food severed from its ecological origins.

Television similarly mobilises tastelessness to reveal environmental harm. In Hulu's *The Handmaid's Tale* (2017–), Gilead's infertility crisis is linked to toxic waste, chemical drift, and ecological collapse. Vegetables, grains, and bread appear bruised, thin, or flavourless. Commanders' households access rare, vibrant produce where as ordinary citizens encounter bland, nutritionally meagre meals. Taste becomes stratified. a sensory marker of who inhabits safe, purified environments and who must live on the residuals of polluted land. The uneven distribution of flavour reveals the unequal distribution of environmental harm. Post-apocalyptic narratives similarly use tasteless food to emphasise environmental ruin. In Albert and Allen Hughes's *The Book of Eli* (2010) food is expired, canned, or nutritionally empty, and characters remark on the dullness or bitterness of what little remains edible. Taste is a casualty of nuclear ruin, the earth no longer produces anything fresh, rendering flavour an archaeological memory rather than a living sensory experience. Tastelessness operates as a critique of toxic landscapes, chemically compromised soil, industrial agriculture, and ecological collapse. It signals that the environment has lost its capacity to sustain flavour, vitality, and multispecies life. Cinema and television thus mobilise taste, its reduction, absence, and artificial reconstruction, as a powerful form of ecological critique. Tastelessness becomes a sensorial index of environmental harm, a sign that toxic residue has entered the most intimate circuits of life. By drawing attention to the mouth as a site of environmental knowledge, these works remind viewers that contamination is not only seen, heard or inhaled but it is eaten, swallowed, and felt.

Toxic Modernity

Contaminated ground is not an anomaly at the edges of modernity, it is one of its most enduring products. Toxic landscapes must be understood as central ecologies of toxic modernity, spaces where industrialism, colonialism, and capitalist extraction have left long, often irreversible, inscriptions in soil, air, water, and flesh. To walk through these landscapes is to move across historical strata of harm. Contamination is both a chemical condition and a cultural one: a set of residues and atmospheres that reorganise how environments are inhabited, sensed, remembered, and contested. By tracking toxicity through atmosphere, residue, and time, contaminated land troubles any tidy separation between ecology and history. As atmosphere, toxicity saturates the air with smog, dust, radiation, spores, and odours, making breath precarious and embedding risk in the everyday act of inhalation. As residue, it congeals in ground, tailings, ruins, household dust, and infrastructural decay, forming geologies of harm that resist remediation and turn landscapes into archives of corporate and state neglect. As time, toxicity unfolds slowly and unevenly, producing illnesses decades after exposure, mutating across generations, and exceeding the temporal horizons of law and policy. Rob Nixon's slow violence provides a crucial frame here: contaminated ground embodies violence that is incremental and attritional rather than explosive, and all the more pervasive because it rarely conforms to event-based notions of disaster.

Contaminated ground is not only a symptom of toxic modernity but a diagnostic of its logics. To attend to contaminated ground is to reckon with the material afterlives of extraction and to confront the fact that environmental crisis is not a rupture but the predictable outcome of centuries of exploitative land use and resource hunger. It is also, however, a threshold. If this chapter has focused on the after-effects of harm, the atmospheres, residues, and temporalities that define contaminated landscapes, the next chapter turns upstream to the cultures that produce such damage in the first place. Extraction, mining, and petro-culture do not simply scar the land, they organise economies, identities, imaginaries, and aesthetic forms. They generate the infrastructures, industries, and everyday practices that will later harden into toxic residues. The coal seam, the open-cut mine, the oilfield, the pipeline, and the refinery are not just technical sites but cultural ones: they are settings for stories, myths of progress, masculinised labour imaginaries, and fantasies of inexhaustible energy creating contaminated landscapes.

References

Australian Federal Government. (2014). *Interim report: Investigation into the environmental impacts of the leach tank failure at Ranger uranium mine* (Report 206). Australian Government Department of Climate Change, Energy, the Environment and Water. https://www.dcceew.gov.au/science-research/supervising-scientist/publications/ssr/interim-leach-tank-failure-report (DCCEEW).

Beck, U. (1992). *Risk society: Towards a new modernity* (M. Ritter, Trans.). Sage.

Braidotti, R. (2013). *The posthuman*. Polity Press.

Brown, K. (2013). *Plutopia: Nuclear families, atomic cities, and the Great Soviet and American plutonium disasters*. Oxford University Press.

Carson, R. (1962). *Silent spring*. Houghton Mifflin.

Flyn, C. (2021). *Islands of abandonment: Life in the post-human landscape*. William Collins.

Mudd, G. M. (2008). The rum jungle U–Cu project: A critical evaluation of environmental monitoring and rehabilitation success. In A. Fourie, M. Tibbett, & J. Wiertz (Eds.), *Proceedings of the 6th International Conference on Mine Closure* (pp. 465–478). Australian Centre for Geomechanics. https://doi.org/10.1007/978-3-540-87746-2_38

Nixon, R. (2011). *Slow violence and the environmentalism of the poor*. Harvard University Press.

Parliament of Australia. (2014). *Annual Reports (No. 1 of 2014)*. Senate Economics References Committee. Canberra: Parliament of Australia. https://www.aph.gov.au/Parliamentary_Business/Committees/Senate/Economics/Annual_Reports/2014/report1/~/media/Committees/Senate/committee/economics_ctte/annual/2014/report1/report.pdf (Australian Parliament House).

Petryna, A. (2004). Biological citizenship: The science and politics of chernobyl-exposed populations. *Osiris, 19*, 250–265. https://doi.org/10.1086/649405

Pulido, L. (2016). Flint, environmental racism, and racial capitalism. *Capitalism, Nature, Socialism, 27*(3), 1–16. https://doi.org/10.1080/10455752.2016.1213013

Schönbohm, A., & Hermann, I. (2024). Dancing with Biopunks: Innovation in the life science industry and science fiction. *In Innovation in Life Sciences: The Digital Revolution*, pp. 3–14. Springer Nature Switzerland.

Schuller, W. K. (2024). Solarpunk: Radical optimism as Praxis. In *Routledge Handbook of Climate Change and Society*, pp. 497–504. Routledge.

Tsing, A. (2012). Contaminated diversity in "Slow Disturbance": Potential collaborators for a livable earth. *RCC Perspectives, 9*, 95–97. https://www.environmentandsociety.org/sites/default/files/seiten_aus_layout_issue_9_final_web.pdf. Environment & Society Portal.

Vint, S. (2007). *Bodies of tomorrow: Technology, subjectivity, science fiction*. University of Toronto Press.

CHAPTER 6

Extraction Landscapes and Petro-cultures

Abstract Modern landscapes are increasingly shaped by contamination rather than exception, as the legacies of industrial production, colonial dispossession, and extractive capitalism have inscribed enduring harm into land, water, infrastructure, and bodies. Toxic environments are not merely damaged sites but material records of political and economic history, storing the residues of development, state violence, and regulatory failure while transmitting injury across generations through heavy metals, radioactive isotopes, petrochemical by-products, and agricultural toxins. Because contamination is often invisible, delayed, and spatially dispersed, its violence rarely appears as a singular event. Instead, it operates as slow violence: an attritional form of harm that accumulates through sediment, groundwater, cellular damage, and inherited vulnerability, while evading spectacle, public attention, and legal resolution. Exposure to toxicity is unevenly distributed, following the spatial logics of colonisation, racial capitalism, and state abandonment, with hazardous industries and waste infrastructures disproportionately imposed on Indigenous lands and marginalised communities. In this sense, toxicity functions as a mode of governance, organising landscapes and lives through managed exposure and environmental precarity. The analysis is organised around three interlinked dimensions: atmosphere, residue, and time. Atmosphere frames toxicity as a sensorial and affective condition, where smog, odour, haze, silence, and particulate air reshape how

J. Coghlan and H. Nolan, *Landscapes That Speak*,
https://doi.org/10.1007/978-3-032-21757-8_6

environments are felt and inhabited. Residue captures the material persistence of harm in tailings ponds, derelict infrastructures, domestic dust, and poisoned soils, producing damaged ecologies that endure long after industries withdraw. Time names toxicity's latency and afterlife, as illnesses emerge decades after exposure, remediation remains incomplete, and closure is continually deferred. Film and television render these processes perceptible through colour, diffusion, soundscapes, and durational form, revealing toxic landscapes as contested terrains of memory, inequality, and ongoing struggle.

Keywords Contamination · Toxicity · Latency · Residue · Sacrifice zones · Environmental injustice · Extractive modernity

Extraction landscapes are not only sites of resource removal, they are spatial expressions of political economy, cultural ideology, and energetic desire. They are distinct landscape forms produced through the material demands of energy capitalism and the imaginaries that legitimise it (Appel et al., 2015; Bridge, 2009). Within these zones, land is reclassified as commodity, labour is recalibrated as managed risk, and nature is rendered legible as resource. Extraction is thus simultaneously material, symbolic, and affective: it reorganises terrain while reshaping how societies value land, normalise sacrifice, and imagine progress. Bridge's analysis positions extractive landscapes as territorial technologies of capital, created through enclosure, measurement, and valuation processes that fracture existing ecological relations in order to produce geologies of value (Bridge, 2009, 1220–1224). Pits, wells, pipelines, and energy corridors function as spatial devices that translate place into abstractions of volume, flow, and yield, enabling incorporation into global markets. These transformations are inseparable from power. Energy infrastructures generate political landscapes of dependence and control, extending far beyond extraction sites through logistical networks that distribute both energy and exposure (Mitchell, 2011).

Petro-culture embeds these infrastructures in everyday life. LeMenager describes oil as an infrastructure of feeling that binds prosperity, mobility, and modern identity to fossil fuels, making dependence appear natural and desirable (LeMenager, 2014). Aesthetic regimes such as the *petro-sublime* stabilise this logic by rendering extraction spectacular and masking its violence through scale and technological grandeur (LeMenager, 2014; Szeman, 2017). As a result, extraction landscapes become iconic scenes of national progress even as they normalise environmental destruction. These landscapes are also organised through sacrifice. Hazardous industries are disproportionately sited on Indigenous and marginalised lands, producing sacrifice zones where harm is rendered politically acceptable (Lerner, 2010; Nixon, 2011; Whyte, 2018). Extraction landscapes therefore function as temporal as well as spatial infrastructures, distributing harm through delay, afterlife, and intergenerational exposure. Under *petro-modernity*, extraction becomes a world-making project: it generates legal, technical, and social orders that exceed the site itself and govern land, labour, and futures. Tsing's work shows the ecological consequences of this process, as extractive zones become landscapes of abandonment where multispecies life attempts to persist amid disturbance (Tsing, 2015). Extraction landscapes thus expose the core logic of energy capitalism, not only to extract value, but to determine which lands, lives, and futures are expendable.

Extraction also produces a particular way of seeing landscape. Cosgrove's argument that landscape is fundamentally a visual and ideological formation helps explain how deserts, mountains, plains, and rivers are reframed as resources awaiting exploitation (Cosgrove, 1984). Deserts appear empty, mountains become ore bodies, and rivers are reimagined as hydropower potential, all transformations rooted in colonial histories that cast land as vacant or underutilised. Jameson's account of late capitalism further clarifies how extractive infrastructures operate as monumental artefacts of their historical moment, expressing expansion, acceleration, and domination over geological and ecological bodies (Jameson, 1991). Harvey's notion of the spatial fix shows how extraction anchors capital through massive infrastructural investment in roads, ports, camps, and grids often accelerating environmental degradation while legitimating regional abandonment (Harvey, 2001). Extraction landscapes also function as thresholds: material and symbolic zones where land is remade into resource. They mark the interface between surface and subsurface, ecological life and geontological matter, and Indigenous Country and settler-colonial property regimes (Yusoff, 2018). Hence, extraction landscapes are liminal environments in which terrain crosses from habitat to asset and from relational land to industrial abstraction. Seen this way, extraction is not an isolated technical activity but a cultural project that encodes modernity's values of growth, dominance, and entitlement into the land itself. The contamination and degradation that follow are not accidental outcomes but structurally predictable consequences. Extractive landscapes are cultural forms that prepare land for sacrifice.

Terrains of Extraction: Mining

Mining is among the most forceful ways humans reshape terrain, transforming landscapes through the removal, displacement, and reconfiguration of earth itself. Open-cut pits, terraced benches, spoil heaps, haul roads, and tailings ponds are not incidental features of industrial activity, they are monumental earthworks that recompose the land's very geometry. In contrast to subterranean extraction, where the violence remains largely out of sight, open-cut mining exposes the act of removal, creating vast voids and stepped amphitheatres in which geology becomes visible as a sequence of scarred strata. They also signify a profound shift in how land is perceived and valued (Arboleda, 2020). Where Country is understood relationally in Indigenous epistemologies, or even aesthetically in pastoral imaginaries, open-cut mines insist on the land as a resource body to be partitioned, blasted, and relocated (Li, 2014).

Mining landscapes also draw attention to the interface between geology and political economy. Extraction forces the earth into a new relation with capital. Its geological strata become the raw inputs for global economies (Yusoff, 2018). Open-cut pits operate as cross-sections of Anthropocene temporality, collapsing millennia of sedimentation into an exposed commodity surface. In these spaces, the earth's deep history is forcibly rewritten into economic time, remade as a ledger of future value or exhausted potential. The pit wall becomes a visible archive of planetary history, repurposed as a commercial inventory. Such geologic violence exposes the ways modernity converts deep time into industrial time, transforming the Earth from a living assemblage of relations into a resource-body partitioned for extraction. Mining landscapes are also inseparable from settler-colonial spatial regimes, which reclassify Indigenous Country as vacant land. The cadastral grid, pastoral lease systems, and mineral tenements overlay existing relational geographies, effacing Indigenous sovereignty to render land extractable. Open-cut mining dramatises this process physically: Country is literally carved open along the same logics that once underpinned frontier expansion. These pits and spoil heaps materialise the settler-colonial assumption that land is divisible, ownable, and infinitely transformable. This logic is central in Ivan Sen's where the mine's expansion erases boundaries between property regimes, ecological worlds, and cultural responsibilities, turning Country into a spacetime of disruption rather than continuity.

Goldstone vividly illustrates how mining operates not only as a process of earth-work but also as a cultural and political order that fundamentally reorganises the landscape. Set on the edges of a mining frontier in remote Queensland, the film situates its narrative within a landscape driven by extraction, literally and symbolically. The open country surrounding the town is marked by drill pads, dust plumes, cleared scrub, and the visual detritus of industrial expansion. These features place the viewer at the threshold between Aboriginal Country and the encroaching forces of mining capital. Sen reinforces this through aerial shots that juxtapose ancestral land with grids of industrial incursion, mapping the tensions between custodial belonging and corporate entitlement. Goldstone itself becomes a layered archive: beneath the mine lies Country; beneath the township lie histories of displacement; beneath the narrative lie cultural claims, environmental wounds, and the sedimented layers of frontier violence. The mining landscape is not simply a backdrop to corruption and exploitation, it is an active participant, structuring the forms of power and dispossession that unfold (Derry, 2016).

Mystery Road: Origin (2022) visualises mining as a creeping frontier not only through its landscape imagery but through the narrative architecture of its plot. Set in the early career of Indigenous detective Jay Swan, the series entwines criminal investigation with the encroachment of a major mining operation on the fictional outback town of Jardine. As Jay investigates the disappearance of a local teen and the suspicious death of a young Indigenous woman, the narrative steadily reveals the mine as a shadow jurisdiction; an economic and political force whose presence shapes the movement of people, the distribution of authority, and the patterns of harm. The plot repeatedly brings Jay into conflict with mining security staff, corporate representatives, and local police who have been drawn into extractive patronage networks. This tension foregrounds how mining companies operate as parallel sovereignties in remote Australia: hiring private security, dictating access to land, influencing policing priorities, and shaping the terms on which Indigenous communities may speak or be heard. The mine's FIFO workforce compounds this dynamic. Characters such as young engineer Mary Allen and the transient crews around her embody the rhythms of extractive labour, short-term contracts, rotational shifts, and the social dislocation that accompanies them. Their presence underscores how *FIFO capitalism* reorganises not just land but relationships, aspirations, and community cohesion. Visually, the series' aerial shots linger over vast terraced pits, haul roads, access gates, and the sprawl of worker compounds, mapping the mine as an expanding geometry of enclosure. These images echo the plot's central concern: that the mine is not merely a backdrop but a force reconfiguring the social world Jay is trying to protect. *In Mystery Road: Origin*, the mine is not simply an economic engine but the slow-moving antagonist of the series reshaping terrain, constricting community space, and shadowing every choice the characters make.

Warwick Thornton's *Sweet Country* (2017), although set in the 1920s, anticipates the spatial and ontological violence of later petro-culture extraction. Within its narrative of frontier injustice, prospecting and small-scale mineral seeking appear as proto-extractive incursions. They are early gestures toward the open-cut logic that will later reshape Australia. The colonial desire to appraise, measure, and exploit land is already present in the film's depiction of settlement. The miners, prospectors, and surveyors who drift through the film embody a worldview in which Country is convertible into claim, tenure, and yield, prefiguring the industrial-scale earth-moving that modern mining would later impose. This incipient earth-work is inseparable from the frontier racism structuring the film: Aboriginal characters are displaced, criminalised, or conscripted into labour as settlers impose cadastral order on relational landscapes. In *Sweet Country*, mining emerges not as neutral economic activity but as a continuation of colonial possession, one in which extraction, violence, and property-making coalesce into the same territorial project (Moss-Wellington, 2024).

Global cinema reinforces these themes through different extractive topographies. *The Devil's Miner* (2005), a documentary set in Bolivia's Cerro Rico,[1] depicts a mountain hollowed by five centuries of silver extraction. The terraced tunnels and collapsing caverns form a living

[1] *Cerro Rico* (Rich Mountain) in Potosí, Bolivia, has been mined continuously since the Spanish discovered its silver deposits in 1545. During the colonial period, it became one of the most intensively exploited mineral sites in the world, supplying vast wealth to the Spanish Empire while causing the deaths of hundreds of thousands and possibly millions of Indigenous and enslaved Andean labourers through the *mita* forced-labour system, toxic exposure, and tunnel collapses. Centuries of extraction have left the mountain geologically unstable, riddled with tunnels, and still worked by impoverished miners, including children, under hazardous informal conditions. Today, Cerro Rico stands as a

register of colonial labour violence, while the boys who work there navigate a world governed by both mineral demand and the mountain's syncretic spirit, El Tío. Extraction here becomes a lethal pedagogy, teaching children to survive within geologic wounds. Similarly, *Nayola* (2022) and *Cartas da Guerra* (2016), though centred on the Angolan civil war, position diamond mining as an extractive infrastructure through which conflict, dispossession, and global capital intertwine. Open-cut pits and artisanal dig sites scar the landscape, signalling the intersection of mineral wealth, geopolitical intervention, and necropolitical governance. In both films, the mine is less a workplace than a spatialised violence whose logics ripple through bodies, kinship networks, and ecological futures (Kirsch, 2014). New Zealand's *The Luminaries* (2020) reimagines the nineteenth-century gold rush as a cosmology of extraction, mysticism, and colonial desire. Its coastal diggings, sluice channels, and makeshift camps depict mining as a disassembling of both land and self. The earth is rendered as a volatile ledger of fortune and loss, where every pit is an incision into a world of competing sovereignties, Māori, colonial, and corporate. Television continues this scrutiny of extraction's cultural and environmental footprint. The Norwegian series *Occupied* (2015–19) situates mining and energy extraction within a geopolitical thriller, revealing how resource sovereignty becomes a terrain of conflict. The mountainous landscape, rich in minerals yet politically contested, is visually fractured by tunnels, access roads, and militarised resource zones. Extraction appears as a territorial claim that reorders national identity and foreign policy alike. Mining emerges across these texts as both earth-work and world-making: an activity that sculpts terrain while altering the moral, political, and ontological conditions under which land, people, and futures are imagined.

The Mining Subject

Mining's earth-work is accompanied by an equally forceful cultural reworking of labour, gender, and risk, producing landscapes that sculpt bodies and subjectivities as much as they excavate terrain. Open-cut operations, with their draglines, excavators and haulage trucks, stage a choreography of machine-enhanced masculinity in which endurance,

paradigmatic site of colonial and postcolonial extractivism, where geological depletion, labour precarity, and historical violence intersect.

stoicism, and physical bravery are celebrated as essential virtues. These machines are not just tools, their sheer scale function as an index of masculine power, reinforcing ideals of toughness, self-reliance, and domination over both land and people. The pit becomes a proving ground where workers test themselves against geological resistance, mechanical hazard, and the rhythms of industrial shift-work. Across film, television, and news media, miners are repeatedly elevated into figures of stoic masculinity, communal loyalty, and moral endurance. John Berger's (1960) notion of the occupational hero captures this ideological manoeuvre: structural danger is converted into personal bravery, and systemic violence is softened through sentimental narratives of sacrifice and camaraderie. This romanticisation is not just a cultural trope but a structuring feature of extractive imaginaries. It is through these representations that hazardous labour is transmuted into national virtue, allowing extractive landscapes that are fractured, polluted, and often abandoned to be read as theatres of courage rather than infrastructures of landscape scarring.

Classic cinematic texts established the affective grammar of this romance. John Ford's *How Green Was My Valley* (1941) has long occupied a canonical place in film history, celebrated for its lyrical evocation of community, memory, and loss. Yet the film functions as a paradigmatic example of how mining is romanticised and structurally depoliticised in popular culture. The narrative centres on the Morgan family and the life of a South Wales coal mining village, yet it consistently reframes industrial labour through pastoral nostalgia. Although the film depicts strikes, wage cuts, environmental degradation, and the intrusion of capital, these structural issues are softened through the sentimental voice-over of Huw Morgan, whose idealised recollections displace systemic critique with emotional longing. Ford's treatment exemplifies the persistent British cultural tendency to sentimentalise the industrial working class, translating historically specific forms of exploitation into a universalised tale of communal virtue and moral decency. Despite the film's apparent sympathy for miners, it largely displaces the lived realities of nineteenth- and early twentieth-century coal labour: respiratory illness, mine collapses, child labour, and the lethal combination of overwork and inadequate safety regulations. Accidents in the film appear as tragic interruptions of idyllic life rather than predictable outcomes of extractive capitalism. The narrative thus reconfigures structural harm as melodramatic contingency, something to be grieved but not interrogated. The film's celebrated cinematography of chiaroscuro interiors, mist-laden exteriors, and carefully composed images of communal ritual works to dehistoricise the mining village. This aestheticisation transforms an economically precarious community into an elegiac landscape of belonging. As Imre Szeman (2019) argues, such nostalgic representations of fossil-fuel labour obscure the political economy of extraction by rendering carbon modernity emotionally coherent and culturally desirable.

The film is foundational because it establishes the sentimental repertoire through which subsequent late twentieth-century mining films such as *Brassed Off* (1996), *Billy Elliot* (2000), and *October Sky* (1999) reinforce this sentimental template. Although situated against deindustrialisation and labour struggle, they frame hardship through nostalgia, family melodrama, and masculine stoicism. The mine becomes a crucible of character rather than a node in global commodity chains or a driver of ecological ruin. This logic becomes a little sharper within contemporary disaster narratives. *The 33* (2015), recounting the 2010 Chilean mine collapse,[2] openly acknowledges managerial negligence and safety breaches but still focuses on miners as heroic survivors whose suffering reaffirms the moral legitimacy of extraction. Catastrophe becomes a vehicle for redemption and subterranean ruin serves as a stage upon which courage, faith, and unity displace any sustained critique of deregulation, corporate profiteering, or state oversight. The Australian Beaconsfield mine collapse[3] is similarly insightful. The live-broadcast rescue of Todd Russell and Brant Webb was framed in national media as a miracle, celebrated through tropes of mateship, endurance, and community solidarity. This affective framing overshadowed decades of safety concerns, seismic instability, and regulatory insufficiencies. The physical instability of the mine and its toxic geology was rewritten as a testing ground for character. The damaged landscape became sublime precisely because it

[2] The 2010 Chilean mine collapse occurred on 5 August 2010 at the San José Mine, when a major rockfall sealed 33 miners approximately 700 metres underground. The miners survived for 69 days in an emergency refuge by rationing food and organising strict routines before being rescued one by one in October using the Fénix capsule. The event drew global attention, highlighting both extraordinary human endurance and long-standing failures in mine safety and regulation within extractive industries.

[3] The 2006 Beaconsfield mine collapse in Tasmania, caused by a seismic event that killed one miner and trapped two others underground for fourteen days, was reported through national media as a narrative of miraculous rescue and mateship.

hosted human endurance. Catastrophe, therefore, did not indict extraction, it consecrated it. Reality television amplifies this ideological work. Programmes such as *Gold Rush* and *Yukon Gold* recast extraction as entrepreneurial adventure, aligning miners with rugged individualism and frontier aspiration.

The subterranean has long been a privileged site of cinematic horror, a space where environmental hostility, historical trauma, and bodily vulnerability converge. Few texts illustrate the conceptual range of underground horror as clearly as *My Bloody Valentine* (1981, 2009). Unlike films that romanticise mining labour or present miners as embodiments of noble working-class virtue, this franchise exposes the underground as a space already saturated with violence, claustrophobia, and historical trauma. The mine is not merely a backdrop but an active, malignant environment: a labyrinth of narrow shafts, abandoned equipment, unstable rock, and echoing darkness that transforms the subterranean into a theatre of both industrial ruin and gothic excess. The film's killer, outfitted in a miner's helmet, gas mask, and heavy protective suit, draws directly on the iconography of extraction. What is ordinarily protective labour gear becomes a monstrous exoskeleton and the pickaxe, a tool of work, becomes a weapon. This conversion of industrial labour into slasher imagery foregrounds the *industrial uncanny*: the return of extraction's latent violence in spectral and monstrous form. The killer functions as a revenant of corporate negligence, unsafe working conditions, and past mine collapses and as an embodiment of the occupational injuries, fatalities, and traumas historically linked to coal extraction. In this regard, *My Bloody Valentine* articulates the inverse of the romantic miner myth and instead of sentimentalising labour it materialises mining's long history of bodily risk and regional precarity. The subterranean environment is central to the film's affective power.

The mine's architecture of ventilation ducts, rail tracks, abandoned tools, and flooded chambers renders the underground not simply closed or confining but actively hostile. The darkness is total, sound reverberates unpredictably, and shafts and tunnels fold back on themselves. The film amplifies the mine's uncanniness through its use of deep shadows, enclosed framing, and disorienting spatial geography. The result is a landscape in which the viewer's sense of orientation collapses, mirroring the characters' own entrapment within a human-made underworld. *My Bloody Valentine* positions the mine as a deathscape forged at the intersection of labour, landscape, and industrial history. It confronts viewers with the violence embedded in extractive infrastructures, even as it converts this violence into gothic and slasher aesthetics. The underground is not a site of masculine virtue or community solidarity but is a material reminder of how extraction reshapes both land and bodies. In rendering the mine horrific rather than heroic, the films provide a crucial counterpoint to the sentimental traditions of mining cinema and open a space for understanding extraction's residues as culturally and environmentally monstrous.

Terrains of Extraction: Dams

Hydropower mega-projects exemplify extraction on a grand scale, sometimes crossing national border, and always transforming river systems into engineered infrastructures that reorder land, water, memory, and sovereignty. More than energy installations, dams are totalising landscape interventions. They carve new shorelines, drown cultural geographies, and impose state-led hydrological regimes that profoundly reshape ecological and social life. In this sense, the dam is a threshold, a site where landscapes cross abruptly from one condition to another, where settlement becomes ruin, river becomes reservoir, and home becomes memory. Lixin Fan's *Up the Yangtze* (2007) offers a penetrating account of the Three Gorges Dam[4] as a landscape-making force, documenting the rising water that steadily consumes villages, farmland, and ancestral sites along the Yangtze River. Forced relocations, collapsing houses, and the creeping edge of the reservoir reveal the dam as a project of state modernity in which the landscape is not simply transformed but overwritten. The riverbank becomes an unstable frontier, with what was solid ground one month lies metres beneath the surface the next. These sequences

[4] The Three Gorges Dam, completed in 2006 and fully operational by 2012, is the world's largest hydroelectric project, spanning 2.3 kilometres across the Yangtze River in Hubei Province, China. Designed to generate electricity, control flooding, and improve river navigation, the dam created a 660-kilometre–long reservoir that submerged more than a hundred towns and villages, displaced over 1.2 million people, and inundated archaeological sites, farmland, and ancestral cemeteries. Its construction has been widely criticised for triggering landslides, altering sediment flows, threatening biodiversity, and exacerbating seismic and ecological instability. As one of the most ambitious landscape-engineering projects in modern history, the dam exemplifies state-driven environmental transformation and the social, cultural, and ecological costs of mega-infrastructure.

render visible the dam's dual spatiality, producing progress narratives while consigning entire lifeworlds to submergence.

This radical reordering of terrain is taken up at a global scale in Jennifer Baichwal and Edward Burtynsky's *Watermark* (2013). *Watermark* is a feature-length documentary that examines humanity's reshaping of the world through water extraction, diversion, damming, and industrial use. The film weaves together footage from 20 locations around the globe including China's enormous Xiaolangdi Dam,[5] the leather tanneries of Dhaka, California's irrigated farmlands, and the concrete spillways of the Hoover Dam[6] to reveal the planetary scale of hydrological modification. The film constructs a visual language of the *extractive sublime*, juxtaposing the aesthetic beauty of patterned reservoirs, irrigation grids, and dam infrastructure with their unsettling ecological implications. Rather than follow a single narrative, *Watermark* assembles a series of vignettes that show water as a cultural, economic, and political force. It highlights how rivers are channelled, rerouted, and industrially engineered,

[5] The Xiaolangdi Dam is a major hydro-engineering installation on the Yellow River in Henan Province, China. Completed in 2001 after seven years of construction, it was designed for flood control, sediment management, irrigation, and hydropower generation. The Yellow River carries some of the highest sediment loads of any river in the world, and Xiaolangdi was built as part of a state project to regulate silt deposition through periodic "flushing" operations that scour sediments downstream. The dam created a large reservoir, necessitating the relocation of more than 180,000 residents and the inundation of towns, farmland, and cultural sites. Although promoted as an engineering solution to centuries of Yellow River flooding, Xiaolangdi has attracted criticism for exacerbating ecological degradation, disrupting fisheries, and intensifying the river's long-term hydrological instability. It stands as a key example of China's technocratic, large-scale reengineering of riverine landscapes.

[6] Completed in 1936 on the Colorado River at the border of Nevada and Arizona, Hoover Dam is one of the most iconic infrastructure projects of the twentieth century. Built during the Great Depression by the U.S. Bureau of Reclamation, it was celebrated as a triumph of American engineering, labour, and state-led modernity. The dam created Lake Mead—one of the world's largest artificial reservoirs—which submerged canyons, cultural sites, and riparian ecosystems while enabling large-scale irrigation, hydroelectric power, and urban expansion across the American Southwest. Over 100 workers died during its construction, and the project required the relocation of towns, Indigenous communities, and traditional lands. In recent decades, declining river flows, prolonged drought, and climate change have exposed vast bathtub rings on Lake Mead's receding shoreline, revealing the artificiality and fragility of the hydrological system the dam sustains. Hoover Dam remains both a symbol of the New Deal's infrastructural optimism and a reminder of the ecological and social instability embedded in large-scale water control.

turning natural hydrological systems into instruments of agriculture, manufacturing, megacity growth, and state power. The film's sequences at Xiaolangdi are especially striking, documenting the flushing of millions of tonnes of sediment in dramatic plumes that underscore the violence of technocratic intervention. Throughout, *Watermark* emphasises the paradox that the very infrastructures designed to control water often reveal the vulnerability of landscapes dependent on such control.

If *Watermark* frames the dam as monument, *DamNation* (2014) turns to the dam as ruin-in-waiting. *DamNation*, directed by Travis Rummel and Ben Knight charts the rising movement in the United States to dismantle outdated or ecologically harmful dams. The film argues that many of the nation's approximately 80,000 dams, once celebrated as monuments of progress, modernity, and engineering prowess, have become environmental liabilities, fragmenting river systems, blocking fish migrations, degrading ecosystems, and accumulating silt that compromises structural integrity. Through archival footage, activist interviews, biologists' testimony, and arresting natural imagery, *DamNation* traces the shift in public perception from viewing dams as symbols of national pride to recognising them as infrastructures of ecological decline. Central to the film are dramatic sequences of dam removal, most notably the Elwha and Glines Canyon dams in Washington State,[7] capturing explosive breaching and surges of sediment-laden water. These scenes emphasise restoration as a form of ecological and cultural repair if not justice. The documentary also foregrounds the political tensions surrounding dams, exposing how federal agencies, energy companies, Indigenous nations, and local communities compete over water rights, ecological stewardship, and cultural survival. In showing the renewal of rivers post-removal, *DamNation* reframes dams as ruins-in-waiting, monumental structures whose eventual obsolescence reveals the fragility of the modernist faith in infrastructural mastery. The film positions dam removal as both

[7] Built in 1913 and 1927 without fish passage, the Elwha and Glines Canyon dams on Washington State's Elwha River exemplified the ecological and colonial violence embedded in early U.S. hydropower. Their construction devastated salmon populations central to the Lower Elwha Klallam Tribe, halted sediment flow, and reshaped the river into an industrial conduit. Although both dams were finally removed between 2011 and 2014—the largest dam-removal project in U.S. history—their demolition underscores a belated form of environmental justice. The restoration of salmon and river processes highlights how the original dams functioned as instruments of dispossession, and how their removal represents not triumph but overdue redress for a century of ecological harm and Indigenous erasure.

ecological necessity and moral reckoning, suggesting that the undoing of hydropower infrastructure may be as historically significant as its construction.

The colonial dimension of hydropower's landscape-making is dramatised in John Boorman's *The Emerald Forest* (1985), which focuses on an Amazonian Indigenous community confronted with the construction of a massive hydroelectric dam. As forest is cleared and river systems are diverted, the film situates the dam as a violent threshold between Indigenous sovereignty and state-backed developmentalism. The project brings militarised labour, displacement, and ecological devastation, rendering the river not as a shared life-source but as a resource to be exploited. Through its portrayal of inundated forest and drowned sacred sites, the film underscores hydropower as a form of colonial earth-making, one that overwrites Indigenous relational geographies with extractive regimes of control. The dam becomes both an ecological wound and an epistemic imposition: a replacement of one world with another. The ideological history of dams as national icons is captured in Pare Lorentz's *The River* (1938), a New Deal–era documentary celebrating the Tennessee Valley Authority. Produced as government propaganda, the film frames dams as embodiments of democratic progress, flood control, and national rejuvenation. Sweeping panoramas of newly built spillways and reservoirs portray hydropower as a benevolent remaking of the American landscape. *The River* reveals how dams have long been central to state mythologies, presented as symbols of mastery over nature, guarantors of prosperity, and monuments to collective purpose. Reading this film alongside contemporary critiques reveals the ideological long arc of hydropower: from nation-building triumph to environmental liability.

Terrains of Extraction: Oil

Oil landscapes occupy a distinctive and multifaceted position within contemporary imaginaries of extraction. Unlike the immediately visible disruptions of open-cut coal or the monumental infrastructure of hydroelectric dams, oil's geographies are defined by dispersal, opacity, and infrastructural reach. Stephanie LeMenager identifies this as the *petroleumscape*: a sprawling assemblage of pipelines, deserts, offshore rigs, highways, suburbs, refineries, and financial platforms that structure everyday life even as they remain ideologically concealed. Oil is in this sense is not only a resource or an energy system but a cultural formation that shapes subjectivity, aesthetics, and political possibility. If coal landscapes are read through heaviness, residue, and ruin, oil landscapes signify mobility and futurity. Yet the promise of boundless movement is structurally tethered to sacrifice zones from the Niger Delta to Alberta's tar sands where extraction produces toxic ecologies, dispossession, and slow violence (Nixon, 2011). These terrains are often framed through the extractive sublime, offshore rigs lit against dark seas, grids of flares rising over desert refineries, or aerial views of vast bitumen tailings ponds. Such imagery aestheticises environmental devastation, converting systemic harm into spectacle.

The oil wells of Baku, located around the Absheron Peninsula in present-day Azerbaijan, represent one of the earliest and most transformative extraction landscapes of the modern world. By the late nineteenth century, Baku had become the epicentre of global oil production, accounting for more than half of the world's supply and shaping emergent petro-capitalist infrastructures long before the more widely mythologised oil frontiers of Texas or the Middle East. The fields at Balakhani, Bibi-Heybat, and Sabunchu were marked by dense forests of wooden derricks, mud volcanoes, and makeshift roads that together formed a chaotic, industrialised terrain. Baku's oilfields were central to the formation of early multinational oil corporations: the Nobel Brothers Petroleum Company (1879) and the Rothschild-backed Caspian and Black Sea Oil Company, which pioneered techniques of drilling, pipeline construction, and kerosene refining at unprecedented scale. These developments transformed Baku into one of the first petro-modern industrial hubs, complete with worker barracks, boomtown neighbourhoods, and intense labour exploitation. The landscape itself became iconic. Early photographers and filmmakers, including the Lumière Brothers, captured the Baku fields as a terrain of mechanical movement and perpetual extraction, towering rigs, erupting gushers, and fields of derricks stretching toward the horizon. These images established some of the earliest visual tropes of extractive modernity. Even into the Soviet period, Baku remained symbolically charged. Stalinist industrial films and propaganda celebrated the oil workers as pioneers of socialist progress, tying hydrocarbon extraction directly to national futurity (Longworth, 1998).

George Stevens's *Giant* (1956), long canonised as a sweeping melodrama about land, family, and racial politics in Texas, is equally foundational as a cinematic meditation on oil and the transformation of the American West by petro-capital. The film was based on the novel by Edna Ferber, a Pulitzer Prize–winning American novelist and playwright whose work examined U.S. cultural mythologies around land, mobility, race, and national ambition. Best known for *So Big* (1924), *Show Boat* (1926), *Cimarron* (1929), and *Giant* (1952), Ferber specialised in expansive, multi-generational narratives that interrogated the social and environmental costs of American expansion. Her fiction often centred on regions undergoing rapid transformation specifically frontier towns, river systems, cattle empires, and nascent oil fields, capturing the tensions between mythic frontier individualism and the structural inequalities of race, gender, and capital.

Although cattle ranching appears to structure *Giant's* early iconography, the narrative is propelled by the arrival of oil extraction, which rewrites both the social fabric and the material landscape of the Benedict estate. The film's famous gusher sequence marks this transformation with almost mythic force. Jett Rink, the poor-white ranch hand who ends up Texas's most oil-rich citizen, silhouetted against the erupting column of black petroleum, emerges as an archetype of the petro-frontiersman: a figure who converts geological strata into instant wealth through perseverance and opportunism. The scene dramatises the *petroleum sublime* (LeMenager, 2014) in which the violent upward eruption of oil is rendered both terrifying and euphorically generative. Oil is not merely a substance, it is a world-making force whose extraction inaugurates new regimes of value, status, and temporality. Crucially, *Giant* frames oil not as a natural evolution of Texan prosperity but as a disruptive agent that destabilises established class hierarchies. The Benedicts' generational ranching empire rooted in land, cattle, and patriarchal inheritance faces a new, volatile competitor in Jett Rink's petro-wealth, which is liquid, speculative, and spectacular. Oil accumulation displaces cattle wealth in both scale and cultural authority, signalling a shift from *pastoral capitalism* to *petro-capitalism*.

The film thus captures the emergence of *petro-masculinity*: a form of gendered authority grounded in extraction, industrial risk, and spectacular wealth rather than agricultural stewardship reflecting the arrival of affective extractivism, meaning an industrial extractivist system that is providing pleasure for the privileged while exploiting the less privileged and damaging the ecosystem. At the same time, *Giant* entwines oil's ascent with the film's racial politics. The Benedicts' ranch is built on Mexican labour and Indigenous dispossession, but the rise of petro-wealth amplifies these inequities. Oil infrastructure produces new forms of racialised exclusion, unequal housing, segregated medical care, and precarious labour. The petro-frontier intensifies the contradictions of settler modernity: wealth concentrates at the top while the labouring bodies who make extraction possible remain marginalised, invisibilised, or actively expelled from the narrative of Texan progress (Bebout, 2011). Environmentally, *Giant* registers oil's capacity to transform land into an extractive grid. Pump jacks and derricks puncture the formerly open horizon; access roads divide the rangeland; the pastoral landscape is remade as an industrial one. This transition is visualised through contrasting aerial compositions: early scenes of unbroken plains give way to geometric arrays of machinery, signals that geological time has been forcibly converted into industrial time. The land, previously idealised as a frontier of personal mastery, becomes a resource body partitioned for extraction. By the film's conclusion, the petro-frontier has become the dominant organising force in the Benedicts' world. In this light, *Giant* stands as an early articulation of the United States' shift into full petro-modernity. Its dramatisation of the gusher as spectacle, its attention to racial and class realignment, and its reconfiguration of land as extractive capital make it a foundational film for understanding the cultural politics of oil and the transformation of western landscapes under fossil capitalism.

More recently, Paul Thomas Anderson's *There Will Be Blood* (2007) stands as one of the most incisive cinematic interrogations of extraction, capitalism, and landscape. Loosely inspired by Upton Sinclair's *Oil!* (1927) the film reframes the emergence of the American petroleum industry as both a geological drama and a cultural genesis story. Its protagonist Daniel Plainview, first an itinerant silver prospector and later an oil baron, embodies the violent fusion of resource extraction and national mythology. From its nearly wordless opening sequence, in which Plainview claws minerals from the earth with primitive tools, the film establishes extraction as an existential condition; a labour of solitary violence, bodily sacrifice, and intimate entanglement with terrain. Anderson renders mining and drilling less as technical practices than as modes of being, through which land is reinterpreted as opportunity, wealth, and destiny. The film's landscape is not a passive setting but an active agent in the drama of petro-modernity. Anderson and cinematographer Robert Elswit frame the extractive frontier as an austere, horizontally expansive terrain. It is sun-bleached, rugged, and apparently untouched and waiting to be remade. This is not an aesthetic choice. It echoes the settler-colonial fiction of empty land, a fiction that underwrites Plainview's ability to acquire leases and claim geological potential as his own. When the first oil derricks rise, their verticality cuts violently through the horizontal horizon, marking points of penetration into the earth's subsurface. The derrick functions as a threshold structure. It is both a monument to industrial ambition and a conduit for subsurface geologies rendered valuable only through extraction. The gusher sequence (now a trope of oil cinema) is a spectacular eruption of oil followed by a catastrophic fire, which materialises the land's volatility, as if the earth itself convulses under the force of its transformation. These scenes stage extraction as a confrontation between geology and capital, a struggle in which the land is both resource and resistant force.

Sound intensifies this extractive aesthetics. Jonny Greenwood's dissonant, percussive score renders drilling as sonic violence, imbuing the industrial landscape with tension, instability, and dread. The drilling sequences echo with metallic resonance and rhythmic pounding, amplifying extraction's physical and psychological pressures. Silence, too, is deployed meaningfully, with long stretches of stillness emphasising the emptiness produced by extraction including emptied relationships, emptied landscapes, and emptied interiors. These tensions between noise and silence mirror the oscillations in Plainview's own temperament, revealing extraction not only as industrial process but as inner compulsion and emotional depletion. Plainview himself exemplifies the *petro-subject:* a person whose worldview is shaped by the extractive logics of petroleum capitalism. His ambition aligns with the frontier ethos of individualism, conquest, expansion, and exclusion. Yet as the narrative unfolds, these values devolve into misanthropy, paranoia, and profound isolation. His relationship with his adopted son deteriorates as Plainview's oil empire grows, suggesting that extraction reorganises not only land but kinship, intimacy, and care. The conflict with local preacher pits religious authority against capitalist will, dramatising a broader contest between moral order and industrial ambition. The film's infamous final scene, a confrontation culminating in murder within the cavernous excess of Plainview's private bowling alley, reveals the total collapse of sociality in a world remade by extractive desire.

What makes *There Will Be Blood* central to understanding extractive landscapes is its sustained attention to land as archival terrain. Beneath the oilfields lie unspoken histories of Indigenous dispossession, ecological disruption, and earlier settler frontiers. The film never explicitly depicts these layers, but it relies on their erasure. Plainview's success presumes a landscape already stripped of competing sovereignties, ecological relations, and non-capitalist forms of dwelling. The drilling rigs, pipelines, and company towns that rise across the plains overlay these suppressed histories with new forms of industrial inscription. The landscape becomes a cross-section of temporalities: geological deep time, colonial time, capitalist time. In this sense, the film visualises extraction as both earth-work and world-making, a process that physically reshapes terrain while symbolically constructing a national identity centred on resource mastery, infrastructural power, and the promise of boundless energy. *There Will Be Blood* shows that extraction is not merely an economic activity but a cultural formation that organises space, identity, and imagination. The film demonstrates how the extractive landscape becomes both a material terrain and a mythic script for modernity, and how the forces that gouge the earth eventually carve into the human subject as well.

The Petro-culture Suburb and Subject

Oil's spatial logic also produces what Imre Szeman (2019) describes as the embodied petroleum subject: modern life including consumer goods, mobility, labour rhythms, suburban infrastructure is materially organised around petro-capitalism. The suburb, the freeway, the plastics industry, and global logistics chains are not merely by-products of oil economies, they are cultural landscapes in their own right, binding citizens affectively to the very systems that imperil them. *Petro-suburbia* is one of the most pervasive but least interrogated terrains of extraction. Unlike coal pits or offshore rigs, suburban landscapes hide their dependence on oil and petrochemicals beneath an aesthetic of ordinariness: suburban streets, driveways, garage doors, retail plazas, and chemically managed lawns. Yet these environments are built from, and maintained through massive petrochemical inputs. Suburbia is not simply where oil is consumed, it is a spatial formation designed through oil (LeMenager, 2014). Automobility, dispersed housing tracts, asphalted road networks, big-box retail, and the logistics corridors that feed them constitute the everyday infrastructures of fossil modernity.

Film and television frequently capture this suburban petro-logic, often unwittingly. *Edward Scissorhands* (1990) presents a pastel-coloured suburb whose manicured lawns, repetitive housing stock, and incessant car trips reveal a world underwritten by petroleum. Its aesthetic uniformity masks *petrochemical saturation* of vinyl siding, fertilisers, plastics, and synthetics that form the material grammar of domestic life. The film's uncanny suburbia subtly registers the disquiet of petro-dependence and the way fossil modernity constructs landscapes that appear naturalised, timeless, and hygienic, even as they rely on continuous extraction elsewhere. Television's *Stranger Things* (2016–) also reveals the petro-suburban imaginary, though through a retro lens. Its 1980s suburbia of tract housing, drive-in diners, cul-de-sacs, mall carparks, and teenage mobility governed by access to petrol constructs a world deeply dependent on fossil-fuel infrastructures. Even the Hawkins Lab and the logistics of containment and secrecy rely on fleets of cars, vans, and fuel-thirsty machinery. The show's nostalgic rendering inadvertently underscores how fossil modernity is woven into the fabric of suburban American childhood.

Breaking Bad (2008–13) renders Albuquerque's suburban expanses as *petro-logistical space*. Walter White's suburban home, the car-wash, the fast-food franchises, and the desert highways collectively form a built environment shaped by cheap fuel and expansive automobility. The series frequently uses wide shots of suburbs hemmed in by multi-lane roads, emphasising how petro-suburbia spatialises moral choice, movement, and economic desperation. While not explicitly about oil, the show's visual world is a fossil landscape with its rhythms, distances, and moral economies organised by the car, the highway, the desert and the mall. A similar petro-cultural configuration is visible in *Barbie* (2023), which pointedly foregrounds the materiality of plastics, automobility, and suburban idealism (Coghlan et al., 2024). Linimal Barbie Land, with its pastel houses, smooth roadways, convertible cars, and frictionless mobility, operates as a stylised *petro-suburbia*; a world literally and figuratively made of plastic, the most ubiquitous petrochemical product of the modern age. Its dream houses, lawns, cars, and consumer goods form a closed-loop system of fossil-derived materiality, rendering petro-culture cheerful, clean, and consequence-free (Brammer, 2026).

The film repeatedly invokes wide, bright shots of seamless road networks and perfectly spaced homes, emphasising a spatial logic organised around car-based movement, synthetic surfaces, and petro-material abundance (Coghlan & Piatti-Farnell, 2025). *Barbie*'s visual world is a fossil landscape par excellence: its infrastructural coherence depends on plastics, fuels, and the suburban fantasy of infinite consumption. When Barbie enters the 'Real World,' the contrast with Los Angeles, an actual petro-city defined by freeways, sprawl, and automobility reveals the degree to which Barbie Land functions as an idealised petro-suburban imaginary. Rather than parodying suburban excess, the film exposes its underpinnings: a world engineered by petrochemical modernity, whose aesthetics of brightness and perfection mask the extractive regimes that make such landscapes possible. Moreso, Barbie herself is a petro-object: a figure literally made from oil-derived plastics whose material composition anchors her within the broader infrastructures of fossil modernity (Boesenberg, 2024).

Her iconic smooth surfaces, glossy hair, synthetic clothing, and endlessly replaceable accessories are all products of petrochemical manufacturing regimes that convert geological deep time into disposable consumer form. Barbie does not merely live in a petro-landscape, she *is* one, an emblem of petroleum reimagined as femininity, aspiration, and endless consumption. Her plastic ontology quietly reveals the fossil underpinnings of girlhood, domesticity, and suburban fantasy, making her an unlikely but potent symbol of extraction disguised as delight (Coghlan et al., 2024). These cultural texts illuminate petro-suburbia as a landscape of everyday extraction, and petro-suburbia is one of the late twentieth and early twenty-first centuries' most powerful visual regimes. It naturalises fossil-fuel dependency, rendering the infrastructures of extraction invisible and transforming the suburb into a reassuring, consumable environment whose ecological costs remain elsewhere.

Like the mining subject, the petro-culture subject is also those who live within, reproduce, and are shaped by the material and ideological infrastructures of fossil modernity, including rig workers, refinery labourers, and truck drivers. Peter Berg's *Deepwater Horizon* (2016) relies heavily on the valorisation of frontline workers, a narrative strategy that aligns with long-standing cinematic traditions in extractive storytelling. The film constructs the rig hands as embodiments of technical skill, camaraderie, and masculine endurance, figures whose courage is rendered all the more admirable because they labour within a setting defined by volatility and risk (Dixon & Gawley, 2017). This heroisation is not incidental. The elevation of workers into moral protagonists is a recurring trope that transforms extractive labour into a theatre of personal virtue rather than a structurally embedded practice shaped by deregulation, corporate governance, and global energy markets. By centring the narrative on the rig crew's bravery during the explosion, the film displaces attention away from the slower, more dispersed violence of the Deepwater Horizon spill.[8] The blowout emerges as a singular event, a spectacular eruption that can be visually mastered and narratively resolved, rather than part of a broader ecology of harm that unfolded over months as millions of barrels of crude entered the Gulf of Mexico. The cinematic focus on the firestorm, collapsing steel, and last-minute escapes narrows disaster into a moment of crisis experienced by identifiable individuals, overshadowing the long-term ecological catastrophe borne by marine ecosystems, coastal communities, and fisheries. In this sense, *Deepwater Horizon* reproduces what Nixon's (2011) *slow violence*: the ongoing, attritional harm of oil toxicity that resists the temporal and aesthetic conventions of blockbuster cinema.

The elevation of workers into heroic victims also reinforces a familiar petro-cultural logic in which "good" labour is set against "bad" corporate behaviour, thereby offering moral clarity without systemic critique. BP executives become the convenient villains, but the offshore industry itself including the regulatory laxity, the political appetite for high-risk drilling, and the labour regimes that normalise danger is left largely untouched. The workers' suffering becomes a moral shield that protects extraction itself from scrutiny, ensuring that the rig is remembered as a site of heroic sacrifice rather than as evidence of the structural precarity inherent to offshore petro-capitalism. As a result, the film recentres human drama at the expense of environmental realities, allowing the deep-water rig to retain its symbolic status as an industrial frontier rather than a geological threshold where environmental harm is inevitable. *Deepwater Horizon* offers a narrative of redemption through resilience, but at the cost of decentring the far larger story of oceanic devastation and petro-industrial accountability. The heroic worker narrative becomes a cultural mechanism through which oil's ecological violence is muted, sentimentalised,

[8] The Deepwater Horizon disaster began on 20 April 2010, when the offshore drilling rig exploded and sank approximately 50 miles off the Louisiana coast, killing eleven workers and initiating one of the largest marine oil spills in history. Over the 87 days before the well was capped on 15 July, an estimated 4 million barrels of crude oil were released into the Gulf of Mexico. The spill contaminated an extensive area, affecting marine wildlife, deep-sea ecosystems, coastal wetlands, fisheries, and shorelines across multiple U.S. states. In addition to crude oil, the event involved major releases of methane and the widespread application of chemical dispersants, introducing toxic compounds including PAHs, VOCs, and heavy metals into marine and atmospheric systems, posing enduring risks to oceanic life and coastal human communities.

or deferred, enabling extraction to reappear as tragic but necessary rather than structurally unsustainable.

Extraction is not simply something that happens *to* landscape but a project that happens *through* landscape. Extraction produces particular terrains, thresholds, and transformations that do more than facilitate the movement of ore, water, or oil; they recode how land is seen, valued, and inhabited. Open pits, pipelines, dams, offshore grids, petro-suburbs, and logistics corridors are not neutral infrastructures. They are spatial arguments about what land is for, who it belongs to, and which forms of life are expendable in the pursuit of energy and profit. In this sense, extraction landscapes are not the background to toxic modernity but one of its primary instruments. Reading extraction through landscape as terrain, threshold, and transformation clarifies how deeply it reorganises both matter and meaning. As terrain, extraction is a form of earth-writing: it inscribes the logics of accumulation into mountains, rivers, forests, deserts, and coasts, suturing capital into geology. As thresholds, extraction zones mark the interfaces where land is forced across ontological boundaries meaning habitat becomes resource, Country becomes tenement, water becomes storage and flow regimes. As transformations, extractive projects crystallise the particular values of modernity, growth, speed, dominance, entitlement, into the land itself, so that future contamination appears less as aberration than as the predictable after-image of those values. This is why the landscapes examined here are never only technical artefacts, they are cultural forms that stage and stabilise a certain relationship between humans, nonhumans, and earth.

Extraction landscapes are also inseparable from the production of extractive subjects. Romantic miners, stoic rig workers, entrepreneurial prospectors, and seemingly ordinary suburban consumers are all positioned within visual and symbolic regimes that translate structural violence into personal virtue, sacrifice, or necessity. These figures are not incidental to extraction, they are the human faces through which sacrifice zones are made tolerable. By elevating courage, hardship, and resilience, extractive cultures convert ongoing harm into a moral drama, displacing attention from the systems that render some lands, bodies, and futures more disposable than others. The same landscapes that carve pits and dam rivers also carve expectations about what counts as a reasonable cost of modern life.

Extractive landscapes are also landscapes of erasure. They depend upon prior acts of forgetting Indigenous sovereignty, of alternative land relations, of the more-than-human communities that are overwritten by cadastral grids and megaprojects. They make deserts look empty, mountains look dormant, rivers look underutilised, and coastal shelves look available for capture. By aestheticising scale and infrastructural prowess, the extractive sublime converts damage into spectacle, inviting awe where anger or refusal might otherwise be possible. In doing so, extraction landscapes help to naturalise the very conditions that will later be read as unfortunate but unavoidable externalities. Seen via this lens, extraction is not an episode within otherwise benign landscapes. It is a way of world-making that saturates terrain, social relations, and imagination. This extends the book's central claim: that landscape is a key medium through which power organises environmental futures. The contaminated grounds of Chapter 5 do not emerge from nowhere, they are the downstream expression of choices already sedimented into pits, dams, corridors, rigs, and suburbs. To contest toxic modernity, then, is not only to remediate its residues, but to interrupt the landscapes and petro-cultures that prepare land for sacrifice in the first place, to learn to see in these terrains not destiny or necessity, but contingent architectures that could, and must, be otherwise.

References

Appel, H., Arthur, M., & Michael, W. (Eds.). (2015). *Subterranean estates: Life worlds of oil and gas*. Cornell University Press.

Arboleda, M. (2020). *Planetary mine: Territories of extraction under late capitalism*. Verso.

Bebout, L. (2011). Troubling white benevolence: Four takes on a scene from *Giant*. *Melus, 36*(3), 13–36. https://doi.org/10.1353/mel.2011.0038, OUP Academic

Berger, J. (1960). *Permanent red: Essays in seeing*. Methuen.

Boesenberg, E. (2024). Saving the planet with barbie? Ecological perspectives on a plastic doll. *M/C Journal, 27*(3). https://doi.org/10.5204/mcj.3069

Brammer, R. (2026). When I wake up in my own pink world: Liminal spaces in the worlds of Barbie. In *The Barbie phenomenon: Gender, identity, race and sexuality* (pp. 120–130). Routledge.

Bridge, G. (2009). Material worlds: Natural resources, resource geography and the material economy. *Geography Compass, 3*(3), 1217–1244. https://doi.org/10.1111/j.1749-8198.2009.00233.x.

Coghlan, J., Hackett, L. J., & Nolan, H. (2024). Barbie: Imagining and interrogating a popular culture icon. *M/C Journal, 27(3)*. https://doi.org/10.5204/mcj.3072.

Coghlan, J., & Piatti-Farnell, L. (2025). Barbie—What was she made for—The changing face of Barbie in popular culture. In *The Barbie phenomenon: Gender, identity, race and sexuality* (pp. 120–130). Routledge.

Cosgrove, D. E. (1984). *Social formation and symbolic landscape*. Croom Helm.

Derry, K. (2016). Goldstone. *Journal of Religion & Film, 20*(3), 33. https://doi.org/10.32873/uno.dc.jrf.20.03.33, digitalcommons.unomaha.edu

Dixon, S., & Gawley, T. (2017). Crude exploration: Portraying industrial disaster in *Deepwater horizon*, a film directed by Peter berg, 2016. *New Solutions, 27*(2), 264–272. https://doi.org/10.1177/1048291117712546

Harvey, D. (2001). Globalization and the "spatial fix". *Geographische Revue, 3*(2), 23–30. https://d-nb.info/1217929630/34

Jameson, F. (1991). *Postmodernism, or, the cultural logic of late capitalism*. Duke University Press.

Kirsch, S. (2014). *Mining capitalism: The relationship between corporations and their critics*. University of California Press.

LeMenager, S. (2014). *Living oil: Petroleum culture in the American century*. Oxford University Press.

Lerner, S. (2010). *Sacrifice zones: The front lines of toxic chemical exposure in the United States*. MIT Press.

Li, T. (2014). What is land? Assembling a resource for global investment. *Transactions of the Institute of British Geographers, 39*(4), 589–602. https://doi.org/10.1111/tran.12065

Longworth, R. C. (1998). Boomtown Baku. *Bulletin of the Atomic Scientists, 54*(3), 34–38. https://doi.org/10.1080/00963402.1998.11456843

Mitchell, T. (2011). *Carbon democracy: Political power in the age of oil*. Verso.

Moss-Wellington, W. (2024). Warwick Thornton's emotional landscapes: Indigenous cinema and cultural autonomy in Australia. *Film Criticism, 48*(1). https://doi.org/10.3998/fc.5692

Nixon, R. (2011). *Slow violence and the environmentalism of the poor*. Harvard University Press.

Szeman, I. (2017). On the politics of extraction. *Cultural Studies, 31*(2–3), 440–447. https://doi.org/10.1080/09502386.2017.1303436

Szeman, I. (2019). *On Petrocultures: Globalization, culture, and energy*. West Virginia University Press.

Tsing, A. (2015). *The mushroom at the end of the world: On the possibility of life in capitalist ruins*. Princeton University Press.

Whyte, K. (2018). Settler colonialism, ecology, and environmental injustice. *Environment and Society, 9*(1), 125–144. https://doi.org/10.3167/ares.2018.090109

Yusoff, K. (2018). *A billion black Anthropocenes or none*. University of Minnesota Press.

Reading the Land: Memory, Power, and Multispecies Futures—A Concluding Invitation

The land remembers
—*True Detective* (Season 1, 2014)

To live in the twenty-first century is to inhabit a world organised by thresholds. Climatic tipping points, ecological volatility, infrastructural fragility, toxic residues, and political upheavals no longer appear as occasional disruptions to a stable order. They shape the conditions of ordinary life. Increasingly, we live in suspension: between crisis and endurance, collapse and persistence, loss and possibility. The landscapes traced throughout this book—liminal, mythic, wonderlands, toxic, extractive—show that thresholds are not exceptional events at the margins of modernity. They are its prevailing spatial form in the Anthropocene. The task, then, is not only to interpret these transitions, but to develop an ethics capable of meeting them, a threshold ethics grounded in responsibility, grief, and refusal.

Thresholds destabilise the inherited moral language of liberal modernity. Much conventional ethics presumes a world of relative stability in which agents act with coherent intention and outcomes remain at least partially foreseeable. Threshold landscapes expose how fragile those premises are. In nuclear exclusion zones, toxicity unfolds across centuries rather than moments, exceeding the human scale of accountability. Along border deserts and detention archipelagos, sovereignty is exercised not

J. Coghlan and H. Nolan, *Landscapes That Speak*,
https://doi.org/10.1007/978-3-032-21757-8

as protection but as suspension, an apparatus that manages vulnerability through delay, exposure, and abandonment. In petro-suburbs, climate harm is structurally veiled by everyday convenience, distributed through habits of comfort that render dependence ordinary and damage distant. In flooded coastlines and drought-stricken plains, causality disperses across generations, severing the neat temporal link between action and consequence. In such conditions, ethics cannot be grounded in control. It must be grounded in attunement, an account of agency as relational, situated, and responsive to worlds already in motion.

Responsibility in threshold landscapes begins not with mastery, but with listening. Indigenous philosophies articulate this with particular clarity: responsibility emerges through relation rather than dominion. Land is not inert matter awaiting use, but sentient, storied, and communicative with obligations enacted through reciprocal care rather than extraction. In threshold environments, this is not simply a moral preference but a practical necessity. Melting glaciers, poisoned creeks, desert winds, and toxic atmospheres all *speak*, albeit in registers that are dispersed, unfamiliar, and often deliberately ignored. Threshold ethics therefore refuses the abstraction that reduces landscapes to scenery, property, or resource. It calls instead for presence and witnessing, for a willingness to be affected, to recognise that the world addresses us materially as well as symbolically. Haraway's injunction to remain with difficulty, Tsing's insistence on noticing, and Whyte's work on Indigenous experiences of repeated climatic disruption converge on a shared proposition: responsibility begins by attending to worlds that are already wounded, already transforming, and already carrying histories that modernity has tried to bury.

Threshold landscapes also compel grief, because they make loss unavoidable, emotional, and materially specific. Coral bleaching across the Great Barrier Reef, deltas disappearing, mass species die-off, cultural sites submerged under reservoirs, forests burned into skeletal ruins: these are not only environmental degradations but extinctions of relation, the unravellings of the human and more-than-human ties that enable life to be liveable. Grief, in this sense, is not only affective, it is epistemic. It names what modernity habitually disavows. It unsettles developmentalist narratives that frame harm as collateral, temporary, or redeemable through future growth. Grief keeps the wound visible. It resists the speed with which extractive systems normalise disaster and demand that societies move on. Grief discloses relationality: the self is not autonomous but constituted through ties that bind it to others, to places, and to futures it does not control. To grieve landscapes is therefore to acknowledge dependency, to admit that the world we inhabit is composed through relations now fraying. Yet grief in threshold landscapes is also anticipatory, looking forward as much as backward. Berlant's account of *crisis ordinariness* names how people come to inhabit ongoing, low-level catastrophe, where exhaustion and dread replace shock. Threshold ethics must therefore refuse grief as paralysis. It must turn grief into praxis: a way of dwelling with irreversibility without surrendering to abandonment.

If responsibility listens and grief acknowledges, refusal intervenes. Refusal is the ethical posture most urgently required in the Anthropocene because it contests the narratives of inevitability that keep extractive worlds running. Refusal interrupts the smooth reproduction of petroculture, settler colonial entitlement, and capitalist futurity by denying that sacrifice is natural, necessary, or beyond contestation. It protects relations against the logics that render land expendable and communities disposable. In threshold landscapes, refusal takes material form: defending water from contamination; stopping mines on sacred Country; resisting hydropower displacement; refusing the abandonment of toxic communities; demanding repair for drowned and erased geographies; rejecting the state's insistence that some places must be ruined for others to flourish. A threshold ethics, therefore, moves across these three interlinked modalities—responsibility, grief, refusal—to cultivate an ethical life suited to worlds that are unstable, compromised, and still becoming. It recognises that the landscapes we inherit are not tidy or whole. They are partial, damaged, and often broken. Yet it insists that damaged worlds still demand care, attention, and defence.

Contemporary culture is saturated with ruined futures. Climate fiction, apocalypse cinema, extinction discourse, and anthropocene narratives repeatedly stage burned forests, flooded cities, cracked deserts, nuclear wastelands, and petrochemical dead zones. These imaginaries do not emerge from speculation alone; they are drawn from material realities already unfolding. Yet the dominance of catastrophic representation can obscure another temporal mode that is increasingly central to life in damaged environments: maintenance. Dystopian landscapes have long functioned as warnings, but many of the terrains examined in this book; Chernobyl's sarcophagus, post-industrial belts, border ruin zones, collapsing dams, toxic coastlines, water-scarce plains, are not allegories of what might come. They are diagnostics of what is already here. They expose the infrastructures that prop up modernity, the pipelines, dams, refineries, highways, and mines precisely at the moment their instability becomes visible. Texts such as *Threads*, *Chernobyl*, *There Will Be Blood*, *Snowpiercer*, and *Beasts of the Southern Wild* show that dystopian landscapes are not endpoints but processes, shaped by political decisions and cultural desires. The risk, however, is that dystopia can foreclose imagination by presenting crisis only as total collapse, leaving little room to perceive the uneven, ongoing labour through which worlds are held together.

In many landscapes, ruination is not an event but a residency. In Fukushima's evacuated zones, families tend abandoned gardens. In Wittenoom, asbestos fibres circulate long after closure, yet people continue to live in their shadows. In Flint, households manage lead exposure through elaborate filtration regimes. In Louisiana's Cancer Alley, communities build activist networks and mutual aid amid petrochemical saturation. In Maralinga, Anangu peoples sustain cultural practice on poisoned ground. In post-industrial towns, gardens bloom in chemically compromised soil. These sites show that ruins are not only places of loss, they are also sites of persistence. They host feral ecologies and unexpected survival, and they hold human collectives engaged in the daily work of endurance. Ruins become temporal hinges where past harm, present adaptation, and future possibility coexist.

To speak of renewal, however, risks romanticising survival. Ruined landscapes cannot simply be restored to an imagined past. They require new forms of governance, new infrastructures, and new imaginaries adequate to altered conditions. Maintenance is not a denial of crisis but a commitment to inhabiting worlds that cannot be fully repaired. A politics of maintenance therefore asks difficult questions: how do communities continue amid structural neglect? What knowledges circulate in damaged worlds? How do ruins become archives not only of harm but of endurance? What solidarities emerge in shared exposure? Ruined futures are not the end of world-making. They are the terrains where world-making becomes most necessary.

Throughout this book, landscape has emerged not as backdrop but as a medium through which political, ecological, and cultural transformations become perceptible. A situated practice of landscape reading therefore demands more than interpretation of representation. It requires approaching landscape as relation rather than object, as an active assemblage through which power, memory, and more-than-human agencies express themselves. Landscapes reveal the operations of power because they encode its spatial logics. Border walls, pipelines, dam walls, mining benches, toxic exclusion fences, and suburban grids don't just occupy terrain, they inscribe governance onto landscapes. A pipeline is not a neutral conduit but a spatial claim. A desert detention camp is not logistical pragmatism but an architecture of abandonment. A dam that floods ancestral valleys reorders hydrology, memory, sovereignty, and belonging. To read landscapes is therefore to trace the uneven geographies of security and exposure that modernity produces and to recognise how these geographies are lived in bodies and breathed through atmospheres.

A situated practice reads beyond the human, recognising landscapes as multispecies entanglements where human histories intersect with animal migrations, plant ecologies, microbial activity, and atmospheric flows. It attends to material memory: isotopes persisting in soil, fibres lingering in air, residues accumulating in water and flesh, clearances reshaping ecologies across centuries, submerged towns lying intact beneath reservoirs, songlines continuing across Country despite dispossession. Landscapes do not forget. They bear the imprint of what has occurred and what has been suppressed. They are archives—partial, fractured, sometimes disguised—yet insistently readable. And they are sensed. Landscapes impress themselves through breath, smell, sound, heat, texture, movement. Cinema makes this palpable: the metallic taste of radiation in *Chernobyl*; the petrochemical haze of *Blade Runner 2049*; the oppressive quiet of *The Last of Us*; the smothering fog of *Silent Hill*. These registers remind us that environments communicate not only through sight but through atmospheres that touch, infiltrate, and transform landscapes and bodies. To read landscapes is therefore never neutral. It is an ethical and political practice, accountable to histories and communities, attentive to erasures, and responsive to what the land asks of us.

A Closing Invitation

Landscapes teach us that worlds are not fixed, they are continually made and unmade. Every threshold is an opening. Every ruin is a reminder that possibility remains. Every mythic terrain holds teachings for endurance. Every dystopian horizon reveals the necessity of maintenance. Every wonderland discloses the fantasies shaping perception. Every toxic residue compels accountability. Every extractive frontier demands transformation. This book closes by returning to the beginning: landscapes are living terrains of meaning. They shape how we imagine crisis, how we endure loss, how we craft ethics, how we build solidarity, and how we create futures.

The invitation, then, is simple and difficult:

Step into the thresholds
Read the land carefully
Refuse what harms
Repair what can be repaired
Imagine otherwise
And learn, always, from the landscapes that live with us.

Index

J. Coghlan and H. Nolan, *Landscapes That Speak*,
https://doi.org/10.1007/978-3-032-21757-8

Kolibri 360 GmbH
Ferdinand-Jühlke-Straße 7
99095 Erfurt, Deutschland
produktsicherheit@kolibri360.de